D0910114

CHICAGO PUBLIC LIBRARY
HAROLD WASHINGTON LIBRARY CENTER

ADVANCES IN MATHEMATICAL SYSTEMS THEORY

Edited by PRESTON C. HAMMER
Professor of Computer Science
& Mathematics
The Pennsylvania State University

ADVANCES IN MATHEMATICAL SYSTEMS THEORY

THE PENNSYLVANIA STATE
UNIVERSITY PRESS
University Park and London

Standard Book Number 271-73132-X
Library of Congress Catalog Card Number 67-27111

Printed in the United States of America

Designed by Marilyn Shobaken

CONTRIBUTORS

Ranan B. Banerji
Systems Research Center, Case Western Reserve University, Cleveland, Ohio

Preston C. Hammer
Head, Computer Science Department, The Pennsylvania State University, University Park, Pa.

Mihajlo D. Mesarović
Director, Systems Research Center, Case Western Reserve University, Cleveland, Ohio

Alan J. Perlis
Head, Department of Computer Science, Carnegie-Mellon University, Pittsburgh, Pa.

A. Wayne Wymore
Systems Engineering Department, University of Arizona, Tucson, Arizona

CONTENTS

INTRODUCTION:

MATHEMATICS, LANGUAGE, AND SYSTEMS

These times are among the most challenging and the most depressing in the history of man. On the one hand, we begin to glimpse the possibilities of really improving the human condition through strategic use of information being accumulated and through the ability to adapt materials and energy to a variety of needs. On the other hand, never before has there been available so much destructive power—and the power to destroy has not been matched by an equally noticeable increase in wisdom. What has this to do with the theory of systems, which, as such, would seem to have little to do with the general well being of man? The first major impetus to the study of systems arose during World War II and had as its objective more effective military operations. However, even this kind of application is related to the well being of people. What is systems theory about? It is part of man's attempt to organize, stimulate, and control the productivity of people. As a science, systems theory is neutral. In applications, it may be used for the summum bonum or not. However, it may enable better understanding of why organizations succeed or fail, and if we can use the theory well, it may contribute to progress. The contributors to this volume, each in his way, would like to see his work used in improving the lot of humanity.

REAL SYSTEMS AND MODELS OF THEM

Man, for a long time, has sensed the complexity of his environment in contrast to his understanding of it. Our reaction to complexity and mystery is consistently one of simplification and abstraction. Languages, which enable communication among people, are one of the most useful kinds of abstractions from "real" systems. Languages not only enable us to discuss real systems, they also allow human control of other humans and discussion of languages and other abstract systems. The use of a particular language depends on education in the use of that language i.e., on people who understand it. Among the specialized sublanguages, mathematical language plays a special role since some mathematical concepts are common to all languages. In some ways, mathematical language represents an attempt to reduce ambiguity of concepts to a minimum, i.e., to convey precise meanings. The success of various kinds of mathematical systems has been well advertised. It would seem almost ungrateful to point out that only in a few cases can mathematics be successfully applied.

What are the qualities of mathematics which make it useful? First, like a common language, mathematics is in the grasp of people, whereas the real systems it models are not. Thus if I say, "There are five people in this room," the interpretation of "five" is considered simple whereas "people" and "room" are by no means simple if considered as physical systems. Numbers are useful, partly because they convey so little information and because, by common consent, they do not change.

Similarly, the use of geometrical surfaces and curves to represent real surfaces and contours results from an enormous reduction in information to gain simplicity. Thus "the plane" is a more useful construct than a much more detailed collection of surfaces would be.

There is a temptation, in every age, for people to emphasize the depths of man's discoveries. This self-congratulation has its uses, but it is deceptive and therefore dangerous. There is today no good mathematical model of a single cell, to say nothing of a human being. Our knowledge seems necessarily to be superficial. Gross common aspects of systems we may discuss; individual or detailed aspects are beyond reach.

At the present time there are a number of distinct mathematical systems, each one of which plays or might play a role in a general theory of systems. I have already mentioned numbers; algebra arose as a language for computational problems and developed into more abstract and general forms. Geometry, the first branch of mathematics to be systematized, might be thought of these days as a modeling of physical systems. Developing geometries in this sense is a responsibility both of mathematicians and of scientists. The marriage of geometries and algebras effected first by Descartes had an enormous impact on mathematics. The introduction of the concept 0 and the decimal system of computation made arithmetic the most important branch of mathematics by present standards.

The infinitesimal calculus generated by Newton and Leibnitz provided for the first time a continuous logic, i.e., the possibility of dealing with dynamic systems in finite terms. The calculus thus offered the same promise of generality in describing systems. However, even minor systems turned out to be unsolvable in closed form and while there were many developments in analysis and differential equations, the means of using these effectively had to await the advent of electronic computers.

Another component of modeling, probability theory, is useful in deliberately admitting evidence which is not complete. While statistical methods are being used to good account, there is a widespread misinterpretation of their roles. The usefulness of such methods lies in helping extract information for decisions when the data and theory are not precise or complete. However, the interpretation of probability as a "true" model of reality is ridiculous.

Symbolic logic provides yet another resource for systems theorists. The first theory of computation was put forth by Turing in the 1930's. Work on recursive functions by Church, Kleene, and Rosser further formed an established connection between computing and logic. Animation of implicative systems arose in cybernetics, and the neural net models of McCulloch and Pitt led, more or less directly, to what is called automata theory. Shannon introduced information theory in a discretized form, subsequent to his application of Boolean algebras to switching circuits. Combinatory analysis and graph theory were revitalized after 1945. Category theory was generated by MacLane and others and lattice theory and universal algebras were applied to studies of automata.

A brave new world of mathematics seemed about to emerge. However, unification of mathematical systems seemed less possible than before. The split between infinitistic and finitistic mathematics was accentuated by Kemeny and others. For example, Turing machines and automata are regarded as rather general abstract systems. Yet they will not effectively model such a "simple" system as a wheel. Thus such systems are more useful in discussing the languages used in models than they are in producing models.

The systems theorist is thus presented with a smorgasbord of worthy systems to assimilate but he will, according to taste, almost inevitably end up either with an algebraic (finitistic), or a geometric analytic (infinitistic) repast. This brings to mind some observations concerning mathematics. It seems to me that mathematics has failed to maintain contact with reality in many ways. The basic concepts of mathematics, interpretable to everyone, are concealed in contexts which fail to establish their roles. While axiomatics has proved a useful research tool, it has been enshrined as an object of worship. While soundness of argument (i.e., rigor) is a worthy objective of any branch of science, it should at all times be kept in mind that standards of rigor are social standards and that absolute rigor is a phantom. Mathematics is overburdened with poor choices of axiom systems, e.g., for topological spaces, metric spaces, and measure theory. If a mathematical theory of systems is going to be of optimal use, it seems necessary, as part of the development, to reconsider the roles of certain areas.

CONTRIBUTIONS TO SYSTEMS THEORY

The lectures on which this book is based represent the views of four men. Mesarović and Wymore are associated with colleges of engineering and Hammer and Perlis are computer scientists. Mesarović and Wymore present more or less formal theories of systems. Perlis discusses the art of computer languages. Hammer presents aspects of a general theory of spaces. Professor Banerji gives an example of an application of the systems theoretic approach.

The points of view are quite distinctive and any attempt to establish harmony would miss the point. However, I will undertake to give a short summarizing description.

For some years Wymore has been formulating mathematical foundations of systems engineering. His approach would be familiar to automata and sequential machine theorists. However, Wymore is concerned with developing a theory which will apply to engineering systems directly. For this purpose the incapacity, already noted, of automata theory to deal with the real-time problems of engineering makes that theory unpalatable. Wymore avoids this difficulty by admitting functions defined on the time-line as inputs, outputs, and states. Moreover, he chooses to permit functions defined on subsets of the real numbers and hence, if he wishes, he can specialize to automata theory. Thus his work is a generalization of automata theory, made in the interests of embracing phenomena familar to everyone, but excluded from automata theory.

A number of specialists claim that the machine theories can represent anything well enough. This might be true in principle. However, simplicity is one of the requirements of models and infinitistic methods were introduced into mathematics to simplify its use. While differential and integral equations can be very complex, they are far simpler to manage than their finite calculus cousins. The plane might be adequately represented by a finite number of points but who will destroy its homogeneity and simplicity and develop such a theory?

Thus Wymore's theory enables use of analysis, geometry, and topology, whereas these are unnatural in automata theory. Now, granted the desirability of a theory of systems which is associated with the mathematics used to solve problems, it is only necessary to see if Wymore has succeeded in his objectives and to find how far he has been able to develop the theory. It would be difficult, actually, for his theory to be invalid since it is a bona fide generalization of automata theory and sequential machine theory. That is, many questions posed in automata theory are meaningful in Wymore's theory and some of the corresponding techniques carry over.

It is necessary, for example, to discuss equivalence of systems and to use structure-preserving mappings from one system to another or to subsystems. It is necessary to consider decomposition of systems into subsystems to introduce parallel and sequential systems. It is also useful to show how the constructs apply to actual problems. Wymore has addressed himself to these problems. It is not clear to me to what extent he has been able to classify his systems but this is a task which will have many ramifications.

To Wymore, considerations of output are ancillary rather than central to his theory. I have not digested precisely what this stance implies: Both input and output, while not necessary if we envisage an isolated system, seem to be of fundamental interest not only in practice but in theory in dealing with sequential and parallel systems. In any case, the reader who finds this theory to his taste may obtain Wymore's recently published book; a second one on the subject, also by Wymore, is forthcoming.

Mesarović's approach to the theory of systems is not readily summarized although, as he says, it starts from a seemingly simple foundation. Mesarović chooses to start with the concept that a system is a relation. That is, a system is a subset of the Cartesian product of a number (finite or infinite) of abstract sets. Now, while a system is a relation, in Mesarović's treatment, not all relations are systems. In fact, he proceeds by breaking the system into two components, say X and Y, each of which are relations; X is interpreted as a collection of inputs and Y a set of outputs. Both are relations, i.e., subsets of products of factor sets contributing to the system. It then becomes clear that "connections" between inputs and outputs must be established and, unless the "operation" of the system is to be simply a black box, connectors are needed.

At this stage, I find that the initial simplicity seems to have vanished. To provide for states (say to *cause* outputs), Mesarović now brings another object into view, a so-called global state object. It almost appears as if he considers states as ancillary to his theory, but this is not in fact the case. Now Mesarović introduces various binary operations as partial algebras since the operations are not assumed to be defined everywhere. This is done in order to discuss "products" of binary relations in the usual way. Because he has started with such a general base, he must worry about connectivity and functionality as he designates two properties. The latter property is analogous to determinacy in automata theory while connectivity is basically a transitivity condition.

Concern with engineering applications means that time, in some form, is necessary for the theory. Mesarović prefers to use any linear order for a time substitute. This contrasts with Wymore's direct use of the real numbers and a scaled time.

The systems theory of Mesarović is both general and subtle. He and his colleagues have seriously discussed causality in relationship to the theory, and members of Mesarović's group deal in concept formulation, artificial intelligence, and so on. It may be assumed therefore that it will be worthwhile to grasp the reasoning behind this approach if only to see what more specialized systems theories omit.

This brings me to my own portion of this book. Since this part is quite different in aspect than that of the other authors, I must provide an orientation. The other authors, like myself, do not accept the present states of our areas as optimal. However, Wymore and Mesarović do not challenge the basic concepts used from mathematics but adapt these to their own ends. For some years I have been engaged in an effort to find which mathematical concepts are basic and to define them so that they can apply to broader areas. At this time I have gone through the recognized major concepts of point-set topology. I have yet to find one which is best explained in the context of topological spaces. The reason for having picked on topology, in part, was that I found that very few concepts touted as basic in topology would apply directly to my work in computing or numerical analysis. Some of the concepts I have considered are continuity, connectedness, separation, neighborhoods, uniformity, perfectness, approximation, dimension, compactness, metrics, limit, filter, openness, and closedness. I have shown that each of these concepts applies effectively, when suitably generalized, to a much wider range of systems than topology would allow. The essential feature of extension was to require that the concepts apply in discrete as well as in continuous cases.

For my chapters I decided to use three concepts—continuity, filters, and approximation—and to then give a presentation of my first attempt at charting mathematical systems. Each of the concepts treated is vital to systems theory, and filter theory could be treated so as to be a theory of systems. My "Chart of Elemental Mathematics" is a somewhat different approach. It is an attempt to obtain a perspective of mathematical areas. Since such a perspective is still lacking in mathematics education, I believe this effort will be useful if it merely irritates some better-qualified people into doing a more thorough job.

The preceding contributions which I discussed are on a theoretical plane. Perlis discusses computer languages as practical systems. To the uninitiated, it might be felt that there should now be theories of computer languages which could enable the construction of these to taste. However, the computer linguist is faced with the necessity of getting things done, and only to a limited degree can he use formal linguistics, logic, or automata theories to help him in this work. Thus the practicing computer linguist must try to develop insight, try to estimate what is important, and he must compromise among ideals, storage space, time, and cost.

The importance of electronic computing to systems theories has a manifold character. First of all, a computing machine is a system and so is a computer language. Thus computers provide data for systems theorists. They also provide a challenge since theories now extant do not provide the scope or power needed to help improve computer languages.

Perlis describes some of the types of computer language systems, e.g., assemblers, compilers, interpreters, and the kinds of systems these are. The reader with some knowledge of computers can here get a bird's-eye view of the problems faced and sometimes resolved by the computer linguist.

The importance of the data structures admitted is shown in the fact that certain of these structures dominate the construction of computer languages. I find Perlis's organization of data structures to be of particular interest: vectors, lists, tables, etc. Evidently less is understood about control structures, which are assuming more importance in advanced programming, especially for second-level compilers and time-sharing.

Now, in what sense is a computer language a system? Is it possible to describe computer languages in the theories of Wymore or Mesarović? A computer language from one standpoint is simply a system which is put into sequence with a computer to accomplish easier man-to-machine-to-man communications. Thus from the formal point of view a computer language is a system. However, it seems to me that the theory of computer languages is not to be readily modeled in either of the systems theories presented here. The reason this seems to be the case is that computer languages involve a depth of level to which systems theories have not adjusted.

R. B. Banerji's chapter on a systems theoretic approach to decomposition of two-person board games provides an example of the use of suitable abstractions in the context of systems theories which display the necessary compromises between generality and detail. To be useful in a given context, a theory must be general enough to embrace all important features and, within that generality, it must be given specificity by providing appropriate constants. Another way of putting it is that generality displays similarities whereas constraints single out individuality. Since two-person board games are numerous in actuality, Banerji defines a theory general enough to embrace these and proceeds to discuss strategies of play. His results will apply to all such games and are of interest independently for their contribution to artificial intelligence.

1

DISCRETE SYSTEMS AND CONTINUOUS SYSTEMS

A. WAYNE WYMORE

I. DEFINITIONS AND EXAMPLES

Discussion

The basic and long-range purpose of the research presented in this paper is to develop a class of mathematical models adequate to describe, at any desirable level of detail, all interesting engineering phenomena and to develop an associated mathematical theory of the manipulation of such models.*

An example of such a class of models is provided, more or less appropriately, by the theory of differential equations. Many interesting engineering phenomena can be described by a set of differential equations and there is a mathematical theory of the manipulation of differential equations to provide solutions, to answer questions concerning stability and so forth. But differential equations are inadequate to describe all interesting engineering phenomena nor is it possible, in most cases, to achieve a satisfactory level of detail. This is true from the systems engineering point of view, particularly where the engineering phenomena considered are always complicated by many interacting components, some of which are more easily described as continuous, some of which are more easily described as discrete but almost none of which, at a certain level of detail, possess the kind of "smoothness" demanded by models specified by differential equations.

* For a discussion of these problems from a somewhat different point of view, see A. Wayne Wymore, *A Mathematical Theory of Systems Engineering: The Elements*, John Wiley & Sons, Inc., New York, 1967.

The theory of differential equations, however, is so well entrenched in engineering practice that there is a tendency by many engineers to confuse the theory of differential equations with systems theory. For instance, an aspect of this confusion is provided by the use of the term, "time-varying." Among most engineers the term time-varying is used in the following context: many classical engineering phenomena have been described by linear differential equations whose coefficients are either constant or functions of time. If the latter is true of the differential equations describing an engineering phenomenon or system, then the phenomenon or system itself is called time varying. This extension of the use of the term time-varying is inappropriate because it is a simple matter to add one more component to the state vector, say x_{n+1}, and one more differential equation,

$$\frac{dx_{n+1}}{d_\tau}(t)=1, \text{ to the set of differential equations}$$

describing the system, and to make the appropriate substitutions of x_{n+1} wherever t appears as an independent variable in the original equations. The new set of differential equations clearly describes the same "real" system as adequately as the original set of differential equations so the system properties are unchanged, but in the new set of differential equations, time does not appear explicitly on the "right-hand sides" as an independent variable and hence the system is no longer "time-varying." At this point the classical engineer exclaims to the systems theorist, "But you've destroyed the linearity of the differential equations!" This is true but irrelevant from the systems theory point of view. What has been demonstrated is that the term time-varying as used in its classical sense by engineers is a property of differential equations and not of systems.

It is thus clear that the theory of differential equations is not the same as systems theory but it is equally clear that an adequate theory of systems must subsume the models of engineering systems specified by sets of differential equations.

Another example of a class of models useful in describing engineering phenomena and systems is provided by the theory of sequential machines, finite and more general automata, and Turing machines. Many interesting engineering phenomena can be described as sequential machines and there is a mathematical theory of the manipulation of sequential machines to minimize the number of states, to decompose a given machine into submachines and so forth.*

In Ginsburg,** for example, a sequential machine is defined to be a

* See, for example, S. Ginsburg, *An Introduction to Mathematical Machine Theory*, Addison-Wesley, Cambridge, Mass., 1962.

** *Ibid.*, page 5.

5-tuple, $Y=(K,\Sigma,\Delta,\sigma,\lambda)$, where K, Σ, and Δ are finite sets representing, respectively, the set of states of the machine Y, the set of inputs to the machine Y, and the set of outputs from the machine Y, and where δ and λ are functions whose values represent, respectively, the next state of the machine Y and the present output of the machine Y given the present state and present input to Y. Hence δ and λ are functions defined on $K \otimes \Sigma$ with values, respectively, in K and Δ.

This theoretical framework begins to look a little more like a basis for a theory of systems. For one thing, the sequential machine formulation specifies what kind of mathematical object a machine is; a machine is a 5-tuple whose elements are mathematical objects of precise types. With a mathematical definition of a machine, it is possible to begin to prove theorems about machines directly. By ascribing special characteristics to the elements of the 5-tuple, special kinds of machines can be discussed and a precise taxonomy can be established.

But, of course, the theory of sequential machines has the basic limitation that it is precisely that: a theory of sequential machines. It is everywhere assumed that information about the machine only at discrete, integral values of time is interesting or available. This is a great simplification from the theoretical point of view but it limits severely the engineering phenomena that can be modeled within the theory. A generalization is necessary.

Of course, it is easy to fulfill the purposes of model development stated in the first paragraph of this section because of the many ambiguous words used in that statement. For example: What is an adequate mathematical model? What is a desirable level of detail? And when is a phenomenon "described"? What is an interesting engineering phenomenon? And what does it mean to manipulate a mathematical model? To what ends?

Answers to the questions about modeling of engineering phenomena can only be given heuristically, by example, and the weight of the evidence will depend on the strength, experience, and ingenuity of the answerer. A modest amount of such evidence in favor of the class of models to be developed herein will be provided as the discussion progresses. In any final sense, however, such questions are unanswerable and any class of models must be judged by its ultimate usefulness from both the theoretical and practical points of view.

But it is quite clear that to develop a class of models and a theory which subsume both the class of models defined by differential equations and the class of models represented by sequential machines, is an eventual necessity. Furthermore, the class of engineering phenomena which, by fiat herewith, are interesting, includes the classes of phenomena usually studied in computer science and in operations research and the classes of phenomena in which there is formalized interaction between man and machine, between man and man, or between system and society. Thus, the class of mathematical models to be developed must be very general mathematically and very powerful from the modeling point of view.

Hence there will always be at least two points of view appropriate to the development of models. One point of view deals with the adequacy of the class of models as models for engineering and, in particular, for systems engineering purposes, and the other point of view deals with the adequacy of the class of models with respect to the development of a useful and interesting mathematical theory.

Any theory of systems adequate for systems engineering must provide for some particular and special manipulations. For one thing, it must be possible to compare systems structurally in a formal way with respect to intrinsic system performance; that is, the theory must discuss isomorphisms and homomorphisms among systems. It is important, furthermore, to be able to compare systems with respect to extrinsic system performance; that is, the theory must discuss input—output equivalence and alternatives.

One of the most important manipulations of system models for the purposes of systems engineering arises from the analysis aspect of the development of mathematical models. It is desirable to be able to analyze a complex real system into simpler components, develop mathematical models of the components and then model also the interfaces or interconnections among the components. This latter process must be carried on within the theory since it is a manipulation of the models. Hence there must be supporting theory dealing with the question, "When and in what sense, does this process yield a model within the given class?"

The other side of this coin is the synthesis problem: "When can a given complicated model be represented as an interconnection of simpler component models?" Both the analysis and synthesis questions must find their place within any theory of systems claiming to be adequate for systems engineering.

There are certainly other properties desirable for a theory of systems. In this short account of the beginnings of a theory, only the discrete/continuous dichotomy will be discussed in any detail.

Now go back to the sequential machine formulation, $Y=(K,\Sigma,\Delta,\delta,\lambda)$, to seek a generalization with which to model both continuous and discrete systems. The idea of a set of states with which to model the internal workings of a system can be retained but represented by an arbitrary set S. The idea of a set of inputs with which to model the kinds of things in the external environment to which a system responds or which a system will accept as input can be retained, but, again, the set of inputs will be represented by an arbitrary set P. (K and Σ are usually assumed to be finite and/or discrete sets.)

If continuous time is to be considered, however, an additional aspect of input must be considered. In the theory of sequential machines it is tacitly assumed that at any given, integral value of time an input value arrives independent of previous input values and, in fact, the theory of sequential machines provides a process for computing the state of the system given arbitrary, finite sequences of input values assumed to present themselves as input at consecutive, integral time values. But in models accounting for system performance on a continuous time scale, it is important to specify not only the input values but also the way in which the input values are organized in time. For example, each of two models may have as the set of inputs the set of all real numbers between -1 and $+1$ but the one system may be able only to handle step function inputs and the other system may be limited to piecewise sinusoidal inputs.

So, if continuous time is to be considered, it is necessary to give not only a set P of inputs acceptable to the system but also to specify how these inputs may be organized in time, that is, to specify a set F of input functions, functions defined on R,* representing the time scale, with values in P.

In the sequential machine formulation, $Y=(K,\Sigma,\Delta,\delta,\lambda)$, the set Δ and the function λ deal with the specification of output from Y. In the generalization presently being developed it seems wise to represent output from a general system by perfectly arbitrary functions of the state of the system and not to insist that the output set and output function be specified as part of the definition of the system. The wisdom of this omission is more apparent in a discussion of complex couplings of systems where a particular system Z may be coupled into Z_1 by one output function and into Z_2 by another output function.

This leaves only the necessity for generalizing λ, the so-called next-state function. The intuitive interpretation of λ is that if, at time n, the machine Y is in a state x ($x \varepsilon K$) and receives the input s ($s \varepsilon \Sigma$), then the state of the machine Y at time $n+1$ is $\lambda(x,s)$. The function λ is always extended to a function $\bar{\lambda}$ defined on the vector product of the set K and the set $\bar{\Sigma}$ of finite sequences of elements of Σ as follows: if $x \varepsilon K$ and $(s_1, \cdots, s_n) \varepsilon \bar{\Sigma}$, then

$$\begin{aligned}
&\bar{\lambda}(x,(s_0, \cdots, s_n)) \\
&= \lambda(x,s_0) \qquad \text{if } n=0, \\
&= \lambda(\bar{\lambda}(x,(s_0, \cdots, s_{n-1})),s_n) \text{ if } n>0
\end{aligned}$$

The function $\bar{\lambda}$ begins to look a little more like a generalization for the

* See Appendix for specialized notation.

computation of the state of the machine Y at time $n+1$ given an initial (at time equal 0) state x and values of an input function between time 0 and n. If n and $(s_0, \cdots, s_n)$ are held constant, then $\bar{\lambda}(x, s_0 \cdots, s_n))$ as x runs over K, determines a mapping of K into itself which could be interpreted as the state of Y at time $n+1$ given the initial state x and the input sequence $(s_0, \cdots, s_n)$.

It is this last interpretation which is chosen for generalization. So far in the development of a class of models for general systems, the following have been discussed: a set S of states, a set P of inputs and a set F of functions defined on R with values in P. In order to generalize the sequential machine formulation, $Y=(K,\Sigma,\Delta,\delta,\lambda)$, an arbitrary subset T of R, a set M of mappings of S into itself, and a mapping σ of $F \otimes T$ onto M are postulated.

The set T represents the time scale of the model; T might be R^{++}, the set of nonnegative real numbers, to reflect the usual assumption concerning models specified by sets of differential equations provided with initial conditions, or T might be I^{++}, the set of nonnegative integers, to reflect the usual assumption in the framework of sequential machines. For modeling or theoretical purposes, however, T might be a perfectly arbitrary subset of R, a short interval with holes in it, the set, $R^- \cup \{0\}$, of nonpositive real numbers, or the whole set R of real numbers, for example. The only restriction that seems called for is that $0 \varepsilon T$ so that an origin for the measurement of time is provided for in any discussion of experiments on the system or system performance generally.

The set M of mappings of S into itself corresponds to the set of mappings $\bar{\lambda}(\cdot,(s_0, \cdots, s_n))$ of K into itself as n runs over I^{++} and $(s_0, \cdots, s_n)$ runs over $\bar{\Sigma}$.

Thus, it seems reasonable to postulate a mapping σ of $F \otimes T$ onto M so that for each input function f $(f \varepsilon F)$ and time t $(t \varepsilon T)$, $\sigma(f,t) \varepsilon M$, that is, $\sigma(f,t)$ is a mapping of S into itself.* Then, if $x \varepsilon S$, $(\sigma(f,t))(x)$ can be interpreted as the state of the system at time t given the input function f and the initial state x.

In summary, the discussion thus far seems to indicate that a system Z can be defined as a set (or 5-tuple) $\{S,P,F,M,T,\sigma\}$ where S is a set (representing the states of Z), P is a set (representing inputs to Z), F is a set of functions defined on R with values in P (representing acceptable input functions), M is a set of mappings of S into itself (representing the totality of state transition functions), T is a subset of R (representing the time scale over which the system exists or is observable or is definable), σ is a mapping from $F \otimes T$ onto M (representing the way the system changes state as a function of input and time).

To say that a system Z *is* a set $\{S,P,F,M,T,\sigma\}$ as described above, is,

*Throughout this volume $\otimes$ indicates a Cartesian product.

however, too inclusive a definition, for it allows mathematical constructions whose "behavior" transcends what engineering experience and intuition would agree is systemlike behavior. Hence it seems necessary further to restrict the constituents S,P,F,M,T and σ.

One such restriction has already been specified: $0 \varepsilon T$.

Another restriction arises from the interpretation of the symbolism $(\sigma(f,t))(x)$, where $f \varepsilon F$, $t \varepsilon T$, and $x \varepsilon S$, as "the state of the system at time t given the input function f and the initial state x." Under this interpretation, what is $(\sigma(f,0))(x)$? This symbolism must mean the state of the system at time 0 given the initial state x and the input function f. If "initial" is taken to mean time $t=0$, then it must be that $x=(\sigma(f,0))(x)$ or that $\sigma(f,0)$ is the identity mapping, ω, of S into itself. Thus it must be that $\omega \varepsilon M$ and $\sigma(f,0)=\omega$ for every $f \varepsilon F$.

Look now at the set F of input functions. Intuition and consistent interpretation demand some special restrictions here. The set R of real numbers and its subsets (such that to each of which 0 belongs) represent the set of possibilities for the time coordination of system behavior and every experiment on a system or every separate instance of its operation will be assumed to begin at 0. Hence in order to discuss simultaneously two experiments, it is necessary to take into account translations of the origin of the time scale. Let f be an input function, $f \varepsilon F$. Then f is a function defined on R with values in P. Now suppose $r \varepsilon R$ and the origin of the time scale is translated to r. Then f determines a new function g on the new time scale: $g(t)=f(t+r)$ for every $t \varepsilon R$. Since the origin of the time scale should be perfectly arbitrary, and f is a legitimate input function on the original time scale, then g should be a legitimate input function also. The function g will be denoted operationally $f \rightarrow r$ and called "the translation of f by r" or, simply, "f arrow r."

Another important consideration with respect to input functions is the possibility of beginning an experiment on a system with an input function f and continuing, at some point in time, with another input function g. This possibility is provided for in a "segmentation" operation on f and g, denoted $f|g$. Thus, $f|g \varepsilon \mathcal{Q}(R,P)$, the set of all functions defined on R with values in P, and is defined as follows:

$$
\begin{aligned}
&(f|g)(t)\\
&=f(t) \quad \text{if } t \varepsilon R^{-},\\
&=g(t) \quad \text{if } t \varepsilon R^{++}
\end{aligned}
$$

A subset F of $\mathcal{Q}(R,P)$ is an admissible set of input functions with values in P if and only if F is closed under translation and segmentation. That is, a subset F of $\mathcal{Q}(R,P)$ is an admissible set of input functions with values in P if and only if whenever f, $g \varepsilon F$ and $r \varepsilon R$, then $f \rightarrow r \varepsilon F$ and $f|g \varepsilon F$.

Now suppose f is an input function and, for the sake of illustration, that $S=R$, $T=R^{++}$. Assume $x \varepsilon R$ and think of plotting $(\sigma(f,t))(x)$ as a function of t; suppose this graph is represented by Figure 1.

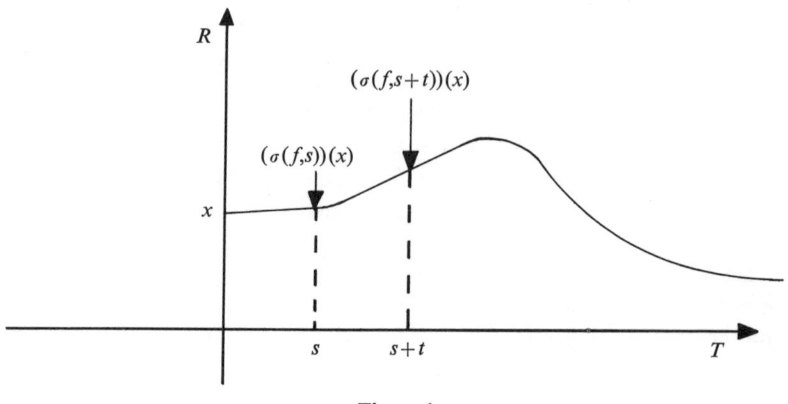

Figure 1

Now assume that $s \varepsilon T$ and $t \varepsilon T$ and imagine that the origin of the time scale is shifted to the point s as in Figure 2.

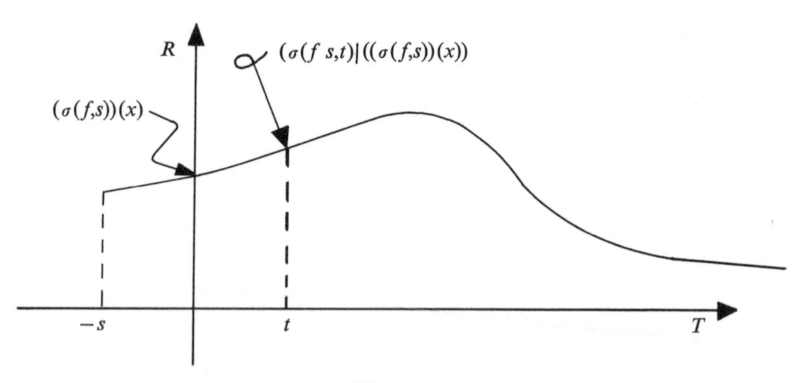

Figure 2

On the new time scale the input function f becomes the input function $f \to s$ and the initial state on the new time scale is $\sigma(f,s)(x)$. At time t, then, the state of the system, by definition, given the input function $f \to s$, and the initial state $\sigma(f,s)(x)$ is $\sigma(f \to s,t)(\sigma(f,s)(x))$. But the only difference between Figures 1 and 2 is a shift in the origin of the time scale and, hence, corresponding points on the state curves should still be the same as indicated by the state x and $\sigma(f,s)(x)$. The only conclusion that can be drawn is that it is necessary to insist that $\sigma(f\rightarrow s,t)(\sigma(f,s)(x))=\sigma(f,s+t)(x)$ if the interpretation of the state of the system is to be consistent.

Another interpretation of this requirement is as follows: suppose the system Z is started in the state x and run till time s with input f. Then the state would be $\sigma(f,s)(x)$. Note that $f\rightarrow s(0)=f(s)$, that is, $f\rightarrow s$ takes up at 0 where f left off at s. Suppose, then, that the system Z is now started in the state $\sigma(f,s)(x)$ with input $f\rightarrow s$ and run till some time t. Then the state would be $\sigma(f\rightarrow s,t)(\sigma(f,s)(x))$. It is postulated that this is the same state that would result in Z if Z were run with input f, from the initial state x, to time $s+t$: $\sigma(f\rightarrow s,t)(\sigma(f,s)(x))=\sigma(f,s+t)(x)$.

Finally, suppose f and g are two input functions for Z which agree on an interval $R(t)$. Then it will be required that $\sigma(f,t)(x)=\sigma(g,t)(x)$ for all $x\in S$, because the system has no way of distinguishing between f and g as far as its input experience is concerned.

These notions are all summarized in the following basic definitions. There are three mathematical constructs involved in the formal definition which follows: the concept of an admissible set of input functions with values in a set P, the concept of an assemblage, and the concept of a system.

Definition 1.1. System Theoretic Definitions. Let P be a set not empty. Then a set F is an admissible set of input functions with values in P if and only if $F\subset\mathscr{Q}(R,P)$, and, for every $f,g\in F$, and $r\in R$, $f\rightarrow r\in F$ and $f|g\in F$ where, for every $t\in R$:

$(f\rightarrow r)(t)=f(r+t)$, and
$(f|g)(t)=f(t)$ if $t\in R^-$,
$\quad =g(t)$ if $t\in R^{++}$

An assemblage is a set $Z=\{S,P,F,M,T,\sigma\}$ where S is a set not empty, P is a set not empty, F is an admissible set of input functions with values in P, $M\subset\mathscr{Q}(S,S)$ such that $\omega\in M$, $T\subset R$ such that $0\in T$, $\sigma\in\mathscr{Q}(F\otimes T$, onto, $M)$.

A system is an assemblage $Z=\{S,P,F,M,T,\sigma\}$ such that

1. for every $f\in F$, $\sigma(f,0)=\omega$;
2. if $f\in F$, s, t, and $s+t\in T$, then $\sigma(f\rightarrow s,t)\sigma(f,s)=\sigma(f,s+t)$;
3. if $t\in T$, f, $g\in F$ and $\mathrm{res}(f,R(t))=\mathrm{res}(g,R(t))$, then $\sigma(f,t)=\sigma(g,t)$

If $Z=\{S,P,F,M,T,\sigma\}$ is a system, $f\in F$, $t\in T$, and $x\in S$, then $\sigma(f,t)(x)$ is called the state of Z at time t given the input function f and the initial state x. If Q is a set not empty and $\zeta\in\mathscr{Q}(S,Q)$, then $\zeta(\sigma(f,t)(x))$ is called the output of Z at time t given the input function f and the initial state

x, with respect to the output function ζ. Functions timetraj $(f,x) \varepsilon \mathcal{Q}(T,S)$, inputtraj$(x,t) \varepsilon \mathcal{Q}(F,S)$ and outputtraj$(f,x,\zeta) \varepsilon \mathcal{Q}(T,Q)$ are defined as follows:

(timetraj$(f,x))(s) = \sigma(f,s)(x)$ for every $s \varepsilon T$
(inputtraj$(x,t))(g) = \sigma(g,t)(x)$ for every $g \varepsilon F$,
(outputtraj$(f,x,\zeta))(s) = \zeta(\sigma(f,s)(x))$ for every $s \varepsilon T$

Discussion

The concept of an admissible set of input functions with values in a set P of input values is used to describe inputs to a system and the way inputs are organized in time. An input function is a function defined on the set R of real numbers with values in an arbitrary set P representing the values (or even, kinds of things) acceptable at the input ports of the system. An admissible set of input functions with values in P is closed under translation and segmentation.

It is clear that $\mathcal{Q}(R,P)$ is itself an admissible set of input functions with values in P and that if F and G are admissible sets of input functions with values in P, then $F \cap G$ is also. If A is an arbitrary family of admissible sets of input functions with values in P then $\cap A$ is an admissible set of input functions with values in P. Consequently, if any subset A of $\mathcal{Q}(R,P)$ is given, then an admissible set of input functions with values in P can be generated. Such an admissible set is denoted $\mathcal{G}(A)$ and is defined as follows: $\mathcal{G}(A) = \cap \{G: G \subset \mathcal{Q}(R,P)$, G is an admissible set of input functions with values in P and $G \subset A\}$. Each element of $\mathcal{G}(A)$ is of the form:

step $(f_0 \to r_0, t_1, f_1 \to r_1, t_2, \cdots, t_n, f_n \to r_n)$ where $n \varepsilon I^{++}$, $\{f_0, \cdots, f_n\} \subset A$, $\{r_0, \cdots, r_n\} \subset R$, $\{t_1, \cdots, t_n\} \subset R$ such that $t_1 < \cdots < t_n$, and step $(f_0 \to r_0, t_1, \cdots, t_n, f_n \to r_n) \varepsilon \mathcal{Q}(R,P)$ is defined as follows: for $t \varepsilon R$,
$(\text{step}(f_0 \to r_0, t_1, \cdots, t_n, f_n \to r_n))(t)$
$= (f_0 \to r_0)(t)$ if $t < t_1$,
$= (f_i \to r_i)(t)$ if $t \varepsilon R[t_i, t_{i+1}]$ for $i \varepsilon I[l, n-1]$,
$= (f_n \to r_n)(t)$ if $t_n \leq t$

Another primitive concept is that of the assemblage. An assemblage has all the constituents of an abstract system but may not exhibit systemlike behavior. Thus, an assemblage is a set Z whose elements are S, P, F, M, T, and σ, that is, $Z = \{S, P, F, M, T, \sigma\}$, where S is a set representing the set of states or internal configurations of the assemblage; P is a set representing the set of inputs to which the assemblage responds; F is an admissible set of input functions with values in P; M is a set of mappings of S into itself representing the total range of behavior of the system in the sense that if the assemblage is in a state x at some time, where $x \varepsilon S$, then the total set of possibilities for the assemblage at some later time is $\{y:y \varepsilon S, y=\alpha(x)$ for some $\alpha \varepsilon M\}$, that is, the total set of possibilities is the set of $\alpha(x)$ as α runs over M; T is the time scale for the system; σ is an onto function defined on $F \otimes T$ with values in M. That is, σ relates the state of the system to input functions and elapsed time so that if x is the state of the assemblage at time 0 and the input function $f \varepsilon F$ is introduced and if $t \varepsilon T$, then $\sigma(f,t)$ is a mapping of S into itself and is an element of M and $\sigma(f,t)(x)$, then, represents the state of the system at time t given the input function f and the initial state x.

The concept of assemblage, however, is too general in that it admits as possibilities mathematical constructs which are pathological from the point of view of representing technological systems and hence the concept of system is defined somewhat more narrowly. A system is an assemblage with three additional conditions.

Because $\sigma(f,t)(x)$ is to be interpreted as the state of Z at time t given the input function f and the initial state x, then $\sigma(f,0)(x)=x$; or, $\sigma(f,0)=\omega$, the identity map for every $f \varepsilon F$.

The requirement that if $f \varepsilon F,s,t,s+t\ T$, and $x \varepsilon S$, then $\sigma(f \rightarrow s,t)(\sigma(f,s)(x))=\sigma(f,s+t)(x)$, is a generalization of the so-called semigroup property described in topological dynamics.*

The requirement that if $t \varepsilon T$ and $f,g \varepsilon F$ such that $\text{res}(f,R(t))=\text{res}(g,R(t))$, then $\sigma(f,t)=\sigma(g,t)$, can be thought of as a time-continuity condition, since $t \notin R(t)$, by definition. Hence, this requirement insists that the instantaneous state of the system at time t does not depend on the instantaneous input at time t but is implied by its input experience up to (but not including) time t.

In order to conserve space here and yet not to exclude all examples, the modeling aspects of the formal systems defined in Definition 1.1 will not be extensively demonstrated, but a few examples of system models will be presented to introduce the theoretical discussion in Section II of this paper.

p \ x	1	2	3
0	2	1	2
1	1	3	2

Table 1

* See, for example, G. D Birkhoff, *Dynamical Systems,* Mathematical Society Colloquium Publication, Vol. 9, 1927.

Consider the finite state machine or automaton represented by the state transition diagram, (Figure 3), or represented equivalently by its next state table, Table 1.

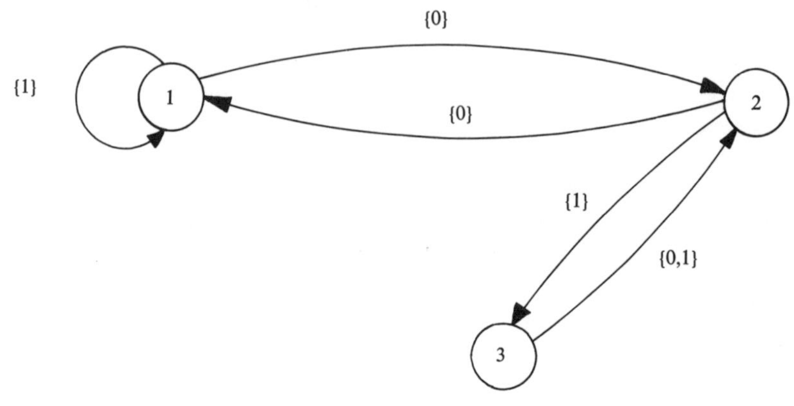

Figure 3

In order to represent this machine in the present theoretical context, it is necessary to define sets S,P,F,M,T,σ such that the behavior of $Z = \{S,P,F,M,T,\sigma\}$ is the same as that of the given finite state machine or automaton and to show that Z is indeed a system according to Definition 1.1.

To this end, let $S = \{1,2,3\}$, $P = \{0,1\}$; let F be the admissible set of input functions generated by the constant functions with values in P; let $M = \text{RANGE}(\sigma)$, $T = I^{++}$; let σ be defined recursively for every $t \varepsilon T$, as follows:

$\sigma(f,0) = \omega$,
$\sigma(f,t) = \sigma(c_{f(t-1)},1)\sigma(f,t-1)$ if $t \varepsilon I^{+}$, where,

for every $p \varepsilon P$, c_p is the constant function on R equal everywhere to p. Then $\sigma(f,t)$ will be completely defined in terms of $\sigma(c_p,1)$ for $p \varepsilon P$:

$\sigma(c_0,1) = \{(1,2),(2,1),(3,2)\}$,
$\sigma(c_1,1) = \{(1,1),(2,3),(3,2)\}$

It is easy to see that the behavior of Z is the same as the original finite state machine. For example, if the input sequence 0,0,1,1,1,0,0, is introduced to the finite state machine and the machine is started in state 2, then it is easy to see from Figure 3 that the finite state machine at time 7, that is, after the sequence has been completely introduced, will be in state 1.

Similarly, if the input function step$(c_0,2,c_1,5,c_0)$ is introduced to the system Z described above and the system is in state 2 at time 0, then

$\sigma(\text{step}(c_0,2,c_1,5,c_0),7)(2)$
$=\sigma(c_{(\text{step}(c0,2,c1,5,c0))(6)},1)\sigma(\text{step}(c_0,2,c_1,5,c_0),6)(2)$
(by the recursion relation),
$=\sigma(c_0,1)\sigma(\text{step}(c_0,2,c_1,5,c_0),6)(2)$
(because $(\text{step}(c_0,2,c_1,5,c_0))(6)=0$),
$=\sigma(c_0,1)\sigma(c_0,1)\sigma(c_1,1)\sigma(c_1,1)\sigma(c_1,1)\sigma(c_0,1)(2)$
(by applying the recursion relation five more times and each time appropriately evaluating $\text{step}(c_0,2,c_1,5,c_0)$),
$=\sigma(c_0,1)\sigma(c_0,1)\sigma(c_1,1)\sigma(c_1,1)\sigma(c_1,1)\sigma(c_0,1)(1)$
(by the definition of $\sigma(c_0,1)(2)$),
$=1$ (by the definitions of $\sigma(c_0,1)$ and $\sigma(c_1,1)$ applied when appropriate)

It can be shown without great difficulty that for arbitrary sets S and P, if F contains the constant functions, if σ is defined recursively by the formulas, $\sigma(f,0)=\omega$ and $\sigma(f,t)=\sigma(c_{f(t-1)},1)\sigma(f,t\text{-}1)$ if $t\,\varepsilon\,I^+$, then, regardless how the mappings $\sigma(c_p,1)$ are defined, on S with values in S, $Z=\{S,P,F,M,T,\sigma\}$ is a system.

Theorem 1.1. Sequential Machines are Systems. Let STATES be a set not empty, let INPUTS be a set not empty and for each $p\,\varepsilon\,$INPUTS let $\alpha_p\,\varepsilon\,\mathcal{Q}(\text{STATES, STATES})$. Let $Z=\{S,P,F,M,T,\sigma\}$ where $S=\text{STATES}$, $P=\text{INPUTS}$, $F=\mathcal{G}(\{c_p\colon p\,\varepsilon\,P\})$, $M=\Gamma(\sigma)$, $T=I^{++}$, and for every $f\,\varepsilon\,F$, $x\,\varepsilon\,S$, $t\,\varepsilon\,T$,
$\sigma(f,t)(x)$
$=x$ if $t=0$,
$=\sigma(c_{f(t-1)},1)(\sigma(f,t-1)(x))$ if $t\neq 0$, where for every $p\,\varepsilon\,P$, $\sigma(c_p,1)=\alpha_p$.
Then Z is a system.

Proof. First of all, Z is an assemblage: $S\neq\emptyset$, $P\neq\emptyset$, F is an admissible set of input functions with values in P, $M\subset\mathcal{Q}(S,S)$ because $\sigma(f,n+1)$
$=\alpha_{f(n)}\cdots\alpha_{f(0)}$ (by induction: $\sigma(f,1)$
$=\sigma(c_{f(0)},1)\sigma(f,0)$ (by the definition of σ),
$=\alpha_{f(0)}$ (by the definition of σ);
if $\sigma(f,n)=\alpha_{f(n-1)}\cdots\alpha_{f(0)}$ then
$\sigma(f,n+1)$
$=\sigma(c_{f(n)},1)\sigma(f,n)$,
$=\alpha_{f(n)}\alpha_{f(n-1)}\cdots\alpha_{f(0)}$), $T\subset R$, $0\,\varepsilon\,T$, and
$\sigma\,\varepsilon\,\mathcal{Q}(F\otimes T,\text{ onto, }M)$ because $M=\Gamma(\sigma)$.

Now $\sigma(f,0)=\omega$ by definition.

Let $f\,\varepsilon\,F$ and $x\,\varepsilon\,S$ and prove that for every s, $t\,\varepsilon\,T$,
$\sigma(f\rightarrow s,t)\sigma(f,s)(x)=\sigma(f,s+t)(x)$. Now

$\sigma(f\rightarrow s,t)\sigma(f,s)(x)$
$=\alpha_{(f\rightarrow s)(t-1)}\cdots\alpha_{f\rightarrow s(0)}(\alpha_{f(s+1)}\cdots\alpha_{f(0)}(x))$ (by the above demonstration),
$=\alpha_{f(s+t-1)}\cdots\alpha_{f(s)}\alpha_{f(s-1)}\cdots\alpha_{f(0)}(x)$,
$=\sigma(f,s+t)(x)$.

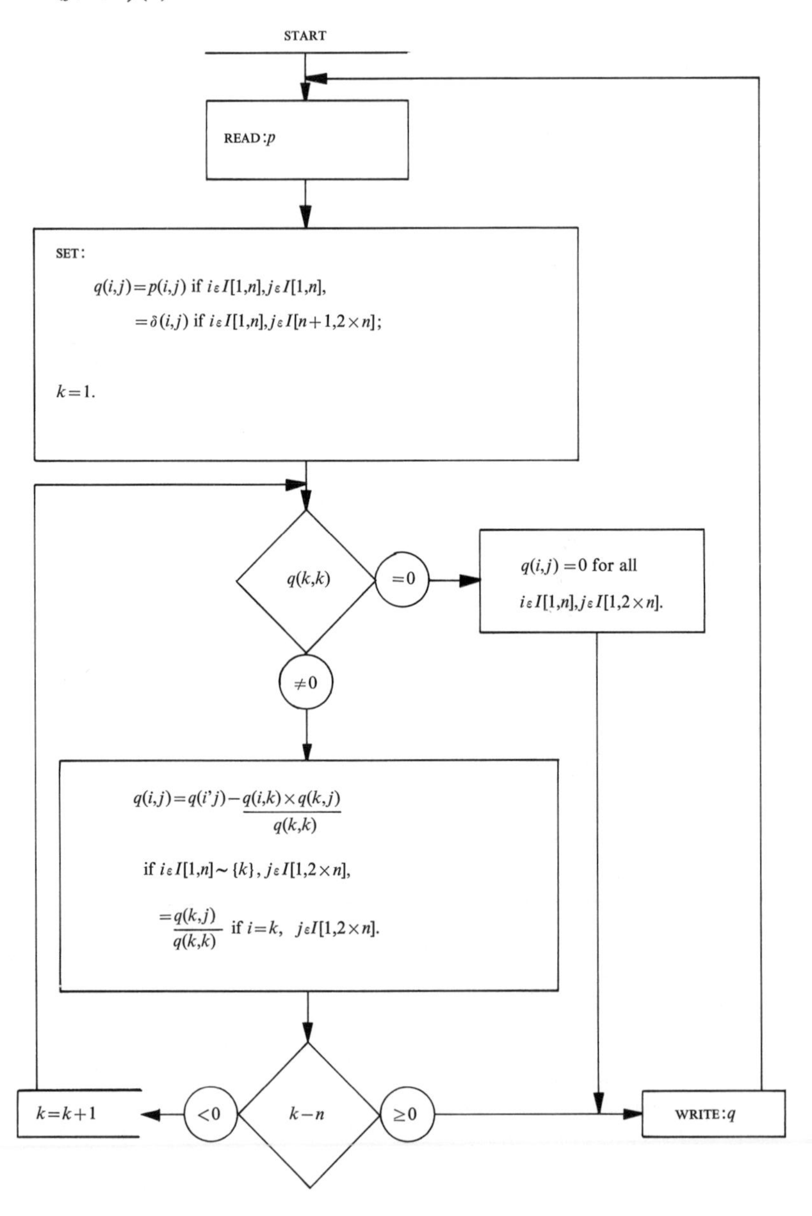

Figure 4

Suppose $f,g \in F$, $t \in T$, $x \in S$ and $\text{res}(f,R(t)) = \text{res}(g,R(t))$. Then

$\sigma(f,t)(x)$
$= \alpha_{f(t-1)} \cdots \alpha_{f(0)}(x)$,
$= \alpha_{g(t-1)} \cdots \alpha_{g(0)}(x)$ (by the assumptions on f and g)
$= \sigma(g,t)(x)$

Hence Z is a system.

Table 2

Conditions \ Components	$\pi_1(\sigma(c_p,1)(x))$	$\pi_2(\sigma(c_p,1)(x))$	$\pi_3(\sigma(c_p,1)(x))$
$\pi_1(x) = \text{READ}$	COMPUTE	1	$(\pi_3(\sigma(c_p,1)(x)))(i,j) = p(i,j)$ if $i \in I[1,n]$, $j \in I[1,n]$, $= \delta(i,j)$ if $i \in I[1,n]$, $j \in I[n+1, 2 \times n]$.
$\pi_1(x) = \text{COMPUTE}$ $\pi_2(x) - n < 0$ $(\pi_3(x))(\pi_2(x), \pi_2(x)) \neq 0$	COMPUTE	$\pi_2(x) + 1$	$(\pi_3(\sigma(c_p,1)(x))(i,j) = (\pi_3(x))(i,j) - \dfrac{((\pi_3(x))(i,\pi_2(x))) \times ((\pi_3(x))(\pi_2(x),j))}{(\pi_3(x))(\pi_2(x),\pi_2(x))}$ if $i \in I[1,n] \sim \{\pi_2(x)\}$, $j \in I[1, 2 \times n]$, $= \dfrac{(\pi_3(x))(\pi_2(x),j)}{(\pi_3(x))(\pi_2(x), \pi_2(x))}$ if $i = \pi_2(x)$, $j \in I[1, 2 \times n]$.
$\pi_1(x) = \text{COMPUTE}$ $\pi_2(x) - n < 0$ $(\pi_3(x))(\pi_2(x), \pi_2(x)) = 0$	WRITE	$\pi_2(x)$	$(\pi_3(\sigma(c_p,1)(x)))(i,j) = 0$ for all $i \in I[1,n]$, $j \in I[1, 2 \times n]$.
$\pi_1(x) = \text{COMPUTE}$ $\pi_2(x) - n \geq 0$	WRITE	$\pi_2(x)$	$\pi_3(x)$
$\pi_1(x) = \text{WRITE}$	READ	$\pi_2(x)$	$\pi_3(x)$

Discussion

Figure 4 is a flow diagram for a digital computer program to invert the n by n matrix p read as input. The flow diagram is a very efficient, very understandable way to represent digital computer programs; it is difficult to improve on the flow diagram as a way to describe computer programs. On the other hand, it is desirable to show that computer programs are not excluded from the theory presently being developed, hence a system model will be defined whose behavior is essentially that of a digital computer under control of the program represented by Figure 4.

To model the flow diagram, let S be the Cartesian or vector product of three sets which are denoted, respectively, as MODES, INDICES, and MATRICES, defined as follows:

MODES = {READ, WRITE, COMPUTE} (that is, MODES is a set whose only elements are the three symbols READ, WRITE, and COMPUTE),

INDICES $= I[1,n]$,

MATRICES $= \mathcal{Q}(I[1,n] \otimes I[1,2\times n],R)$.

Let the set P be the set of real n by n matrices, namely, $\mathcal{Q}(I[1,n]\otimes I[1,n],R)$. Let F be the admissible set of input functions with values in P generated by the constant functions with values in P. The system model selected will be time discrete by defining: $T=I^{++}$, and σ, recursively as follows: $(\sigma f,0)=\omega$, $\sigma(f,t)=\sigma(c_{f(t-1)},1)\,\sigma(f,t-1)$ if $t \varepsilon I^{+}$. Then, Z is a system by Theorem 1.1 no matter how the $\sigma(c_p,1)$ are defined, but the $\sigma(c_p,1)$ have to be defined in such a way that Z will compute the inverse of whatever matrix is input to Z when the MODE component of Z's state is READ.

Table 2 shows how the three components of Z's state must change under different conditions of the present state denoted in the table by x. Here x has three components denoted categorically by $\pi_1(x)$, $\pi_2(x)$, and $\pi_3(x)$, denoting, respectively, x's coordinate in MODES, INDICES, and MATRICES. Note that $\pi_3(x)$ is an n by $2\times n$ matrix. Table 2 can be translated directly into an explicit, but still coordinatewise, definition of $\sigma(c_p,1)$:

$\pi_1(\sigma(c_p,1)(x))$

= COMPUTE	if $\pi_1(x)=$ READ or
	if $\pi_1(x)=$ COMPUTE, $\pi_2(x)<n,(\pi_3(x))(\pi_2(x),\pi_2(x))\neq 0$,
= WRITE	if $\pi_1(x)=$ COMPUTE, $\pi_2(x)<n$, $(\pi_3(x))(\pi_2(x),\pi_2(x))=0$ or
	if $\pi_1(x)=$ COMPUTE, $\pi_2(x)\geq n$,
= READ	if $\pi_1(x)=$ WRITE;

$\pi_2(\sigma(c_p,1)(x))$

$$
\begin{aligned}
&= 1 && \text{if } \pi_1(x) = \text{READ} \\
&= \pi_2(x) + 1 && \text{if } \pi_1(x) = \text{COMPUTE}, \pi_2(x) < n, (\pi_3(x))(\pi_2(x), \\
& && \quad \pi_2(x)) \neq 0, \\
&= \pi_2(x) && \text{otherwise;}
\end{aligned}
$$

$$
\begin{aligned}
&(\pi_3(\sigma(c_p,1)(x)))(i,j) \\
&= p(i,j) && \text{if } i \varepsilon I[1,n], j \varepsilon I[1,n], \pi_1(x) = \text{READ}, \\
&= \delta(i,j) && \text{if } i \varepsilon I[1,n], j \varepsilon I[n+1, 2\times n], \pi_1(x) = \text{READ}, \\
&= (\pi_3(x))(i,j) - \frac{((\pi_3(x))(i,\pi_2(x))) \times ((\pi_3(x))(\pi_2(x),j))}{(\pi_3(x))(\pi_2(x),\pi_2(x))} \\
& && \text{if } i \varepsilon I[1,n] \sim \{\pi_2(x)\}, j \varepsilon I[1,2\times n], \\
& && \pi_1(x) = \text{COMPUTE}, \quad \pi_2(x) < n, \quad (\pi_3(x))(\pi_2(x), \\
& && \quad \pi_2(x)) \neq 0, \\
&= \frac{(\pi_3(x))(\pi_2(x),j)}{(\pi_3(x))(\pi_2(x),\pi_2(x))} && \text{if } i = \pi_2(x), j \varepsilon I[1,2\times n], \pi_1(x) = \text{COMPUTE}, \\
& && \pi_2(x) < n, (\pi_3(x))(\pi_2(x),\pi_2(x)) \neq 0, \\
&= 0 && \text{for all } i \varepsilon I[1,n], j \varepsilon I[1,2\times n] \text{ if} \\
& && \pi_1(x) = \text{COMPUTE}, \\
& && \pi_2(x) < n, (\pi_3(x))(\pi_2(x),\pi_2(x)) = 0, \\
&= \pi_3(x) && \text{otherwise.}
\end{aligned}
$$

It is tedious, but not at all difficult to verify that if $p \varepsilon P$, that is, if p is an n by n matrix, then the MATRICES-coordinate of $\sigma(c_p,n)$(READ, $1,q$), where $q \varepsilon$ MATRICES is arbitrary, is an n by $2\times n$ matrix (formally a function on $I[1,n] \otimes I[1,2\times n]$) such that the matrix consisting of the first n rows and columns is an n by n identity matrix and the last n by n block is the inverse of p—if the inversion procedure doesn't break down. If the inversion procedure breaks down $((\pi_3(x))(\pi_2(x),\pi_2(x)) = 0$, then everything is zeroed out ($\pi_3(x)$ is set to the zero matrix) at the time of breakdown.

To define an output function ζ formally for this system means to define a function on S with values in some set Q. The appropriate output, of course, is the inverse of the input matrix. Hence Q must include the inverses of all nonsingular n by n matrices. If there is no interest in any output except the final result, the machine can output a WAIT! signal till the MODE component of the machine's state turns to WRITE. The symbol which the system emits to indicate it hasn't yet produced the inverse will be the symbol WAIT!. Hence, define Q to be $\{\text{WAIT!}\} \cup \mathcal{Q}(I[1,n] \otimes I[1,n], R)$ and ζ as follows for $x \varepsilon S$, $i,j \varepsilon I[1,n]$:

$\zeta(x) = \text{WAIT!}$ if $\pi_1(x) \neq \text{WRITE}$,

$(\zeta(x))(i,j) = (\pi_3(x))(i,j+n)$ if $\pi_1(x) = \text{WRITE}$.

It is easier, in some ways, to produce examples of real engineering systems in continuous time than it is to produce examples of real discrete systems. In fact, it can be shown that if the vector equation

(1) $\frac{du}{d\tau}(t)=G(u(t),f(t)),\ u(0)=x$

has a unique solution for each f in some admissible set F of input functions with values in a set P, for each x in some vector space S and for each t in some subset of R, and, if $Z=\{S,P,F,M,T,\sigma\}$ where $M=\text{RANGE}(\sigma)$ and $\sigma(f,t)(x)=u(t)$ where u is the solution of Equation 1, then Z is a system.

Theorem 1.2. Differential Equations Define Systems. Let STATES be a Banach space, let INPUTS be a set not empty, let FORCINGFUNCTIONS be an admissible set of input functions with values in INPUTS, let $G\varepsilon\mathcal{F}(\text{STATES}\otimes\text{INPUTS},\ \text{STATES})$. Assume that for every $f\varepsilon$ FORCINGFUNCTIONS and $x\varepsilon$ STATES there exists a unique function $z(f,x)\varepsilon F(R^{++},\ \text{STATES})$ which is the solution function of the differential equation:

$$\frac{dy}{d\tau}(t)=G(y(t),f(t)) \text{ for every } t\varepsilon R^{++},$$

with $y(0)=x$, and such that $\frac{dz(f,x)}{d\tau}$ is continuous almost everywhere.

Let $Z=\{S,P,F,M,T,\sigma\}$ where $S=\text{STATES}$, $P=\text{INPUTS}$, $F=$ FORCINGFUNCTIONS, $M=\Gamma(\sigma)$, $T=R^{++}$, and for every $f\varepsilon F$, $t\varepsilon T$, and $x\varepsilon S$, $\sigma(f,t)(x)=(z(f,x))(t)$ Then Z is a system.

Proof. First of all, prove that if g is any function defined on R^{++} with values in STATES, if r, $t\varepsilon R^{++}$ and g is differentiable at $r+t$, then $g\rightarrow r$ is differentiable at t and $\frac{d(g\rightarrow r)}{d\tau}(t)=\left(\frac{dg}{d\tau}\rightarrow r\right)(t)$. Consider the following computation in support of this assertion:

$$\frac{d(g\rightarrow r)}{d\tau}(t)$$

$$=\lim\{\{(h,\frac{1}{h}\times((g\rightarrow r)(t+h)-(g\rightarrow r)(t))): h\varepsilon R^{+}\},\geq\}$$

(by the definition of the derivative if the limit of the net exists),

$$=\lim\{\{(h,\frac{1}{h}\times(g(r+t+h)-g(r+t))): h\varepsilon R^{+}\},\geq\}$$

(by the definition of $g\rightarrow r$),

$$=\frac{dg}{d\tau}(r+t) \text{ (by the assumption that } g \text{ is differentiable at } r+t),$$

$$=\left(\frac{dg}{d\tau}\rightarrow r\right)(t)$$

Now it's clear that Z is an assemblage since $S\neq\emptyset$, $P\neq\emptyset$, F is an admissible set of input functions with values in P, $M\subset\mathcal{F}(S,S)$, $T\subset R$, $0\varepsilon T$, and $\sigma\varepsilon\mathcal{F}(F\otimes T,\ \text{onto},\ M)$.

Furthermore, for every $f \varepsilon F$, and $x \varepsilon S$, $\sigma(f,0)(x) = z(f,x)(0) = x$ by definition of $z(f,x)$.

Let $f \varepsilon F$, $s,t \varepsilon T$, and $x \varepsilon S$, then,

$\sigma(f \to s,t)(\sigma(f,s)(x))$

$= (z(f \to s, \sigma(f,s)(x)))(t)$ (by the definition of σ),

$= (z(f \to s,(z(f,x))(s)))(t)$ (by the definition of σ),

$= (z(f,x) \to s)(t)$ (by the uniqueness assumption because $z(f,x) \to s$ satisfies the differential equation $\frac{dy}{d_\tau}(r)$ $= G(y(r), f \to s(r))$ with $y(o)$ $= (z(f,x))(s)$: $\frac{d(z(f,x) \to s)}{d_\tau}(r)$ $= \frac{dz(f,x)}{d_\tau}(s+r)$(as in the first paragraph), $= G(z(f,x)(s+r), f(s+r))$ (by the assumptions on $z(f,x)$), $= G((z(f,x) \to s)(r),(f \to s)(r))$, and $(z(f,x) \to s)(0) = (z(f,x))(s)$),

$= (z(f,x))(s+t)$,

$= \sigma(f,s+t)(x)$ (by the definition of σ).

Now assume that $f,g \varepsilon F$, $t \varepsilon T$ and that $\text{res}(f,R(t)) = \text{res}(g,R(t))$. For every $s \varepsilon R^{++}$, let $u(s)$

$= (z(g,x))(s)$ if $s < t$,

$= (z(f,x))(s)$ if $t \leq s$. Then

$\frac{du}{d\tau}(s)$

$= \frac{dz(g,x)}{d\tau}(s)$ if $s < t$,

$= \frac{dz(f,x)}{d\tau}(s)$ if $t \leq s$;

$= G((z(g,x))(s), g(s))$ if $s < t$,

$= G((z(f,x))(s), f(s))$ if $t \leq s$ (by the assumptions on $z(g,x)$ and $z(f,x)$);

$= G(u(s), f(s))$ for all s (by the definition of u and because $\text{res}(f,R(t)) = \text{res}(g,(t))$)).

Furthermore, $u(0) = (z(g,x))(0) = x$. Hence u is a solution of the equation $\frac{dy}{d\tau}(s) = G(y(s), f(s))$ with $y(0) = x$. By the uniqueness assumption $u = z(f,x)$.

And therefore:

$z(g,x)(t) = x + \int_0^t \frac{dz(g,x)}{d\tau'}(\tau)\, d\tau$ (by the uniqueness and continuity assumptions),

$=x+\int_0^t G((z(g,x))(\tau),g(\tau))d\tau$ (because $z(g,x)$ is the solution of the equation $\frac{dy}{d\tau}(s)=G(y(s),g(s))$),

$=x+\int_0^t G(u(\tau),f(\tau))d\tau$ (by the definition of u and the assumptions on f and g),

$=x+\int_0^t G((z(f,x))(\tau),f(\tau))d\tau$ (by the above demonstration),

$=x+\int_0^t \frac{dz(f,x)}{d\tau'}(\tau)\, d\tau,$

$=(z(f,y))(t)$. Hence z is a system.

Discussion

Examples of such systems abound in the engineering literature. Two examples of systems defined by differential equations will be discussed here.

Consider the differential equation:

$$\frac{dy}{d\tau}(t)=\frac{1+y(t)^2}{y(t)}\times f(t) \quad \text{if } y(t)\neq 0,$$

with the initial condition $y(0)=x$, where f is some forcing function. Equation (2) does not necessarily describe any real system so it is possible to assume that the value of $\frac{dy}{d\tau}(t)$ for $y(t)=0$ is not interesting, that is, a function $y(t)$ is desired such that $y(0)=x$ and, if $y(t)\neq 0$, then $\frac{dy}{d\tau}(t)$ is given by Equation (2).

Under these assumptions, the solution of Equation (2) is $y(t)=((1+x^2)\times(\exp(2\times\int_0^t f(\tau)d\tau))-1)^{1/2}$. In order to specify a system, $Z=\{S,P,F,M,T,\sigma\}$, whose behavior is described by the solution of Equation (2), proceed as follows: Let $S=R^{++}$, $P=R^{+}$, $F=\{f{:}f\varepsilon F(R,P)$, $\int_0^t f(\tau)d\tau$ exists for every $t\varepsilon R^{+}\}$, $M=\text{RANGE}(\sigma)$, $T=R^{++}$,

$$\sigma(f,t)(x)=((1+x^2)\times(\exp(2\times\int_0^t f(\tau)d\tau))-1)^{1/2}, \text{for every } f\varepsilon F,\ t\varepsilon T,\ x\varepsilon S.$$

It isn't difficult to show that Z, as defined here, is a system. Show first, in a straightforward manner, that F is an admissible set of input functions with values in P. Next, if $(r)^{1/2}$ is interpreted as the positive square root of r for every $r\varepsilon R^{++}$, and, since P has been restricted to R^{+}, it's clear that

$(1+x^2)\times(\exp(2\times\int_0^t f(\tau)d\tau))\geq 1$ for every $x\varepsilon S$, $t\varepsilon T$ and $f\varepsilon F$, so that

$\sigma(f,t)\varepsilon\mathcal{Q}(S,S)$. This is about all that's necessary to see that Z is an assemblage.

To see that Z is, in fact, a system, check the following deductions for $f,g\,\varepsilon\,F$, $x\,\varepsilon\,S$ and $s,t\,\varepsilon\,T$:

1. $\sigma(f,0)(x)$
   $=(1+x^2-1)^{1/2},$
   $=x;$
2. $\sigma(f\rightarrow s,t)(\sigma(f,s)(t))$
   $=((1+(\sigma(f,s)(x))^2)\times(\exp(2\times\int_0^t f\rightarrow s(\tau)d\tau))-1)^{1/2},$
   $=((1+(1+x^2)\times(\exp(2\times\int_0^s f(\tau)d\tau))-1)\times$
   $(\exp(2\times\int_0^t f\rightarrow s(\tau)d\tau))-1)^{1/2},$
   $=((1+x^2)\times(\exp(2\times\int_0^s f(\tau)d\tau))\times(\exp(2\times\int_0^t f\rightarrow s(\tau)d\tau))-1)^{1/2},$
   $=((1+x^2)\times(\exp(2\times\int_0^s f(\tau)d\tau))\times(\exp(2\times\int_s^{s+t} f(\tau)d\tau))-1)^{1/2},$
   $=((1+x^2)\times(\exp(2\times\int_0^{s+t} f(\tau)d\tau))-1)^{1+2},$
   $=\sigma(f,s+t)(x);$
3. if $\text{res}(f,R(t))=\text{res}(g,R(t))$, then
   $\int_0^t f(\tau)d\tau=\int_0^t g(\tau)d\tau$ and hence
   $\sigma(f,t)=\sigma(g,t)$

Therefore, Z, as defined, is a system.

To illustrate some points discussed earlier, consider, instead of Equation (2), a slight modification as follows:

$$\frac{dy}{d\tau}(t)=\frac{1+y(t)^2}{y(t)}\times\frac{f(t)}{1+t^2}\quad\text{if } y(t)=0 \tag{3}$$

and $y(0)=x$. The solution of this equation is just as easy as it was before:

$$y(t)=((1+x^2)\times(\exp(2\times\int_0^t\frac{f(\tau)}{1+\tau^2}d\tau))-1)^{1/2}$$

and the definition of an assemblage Z could proceed as before. But such an assemblage is not a system because condition 2 breaks down due to the time-varying nature of Equation (3).

To cure this condition, replace Equation (3) with the system of equations:

$$\frac{dy_1}{d\tau}(t)=\frac{1+y_1(t)^2}{y_1(t)}\times\frac{f(t)^2}{1+y_2(t)^2}, \tag{4}$$

$$\frac{dy_2}{d\tau}(t)=1,$$

with $y_1(0)=x_1$, and $y_2(0)=x_2$. Now the solution is:

$$y_1(t)=((1+x_1^2)\times(\exp(2\times\int_0^t \frac{f(\tau)}{(1+(\tau+x_2)^2)}d\tau))-1)^{1/2},$$

$y_2(t)=x_2+t$. The specification of $Z=\{S,P,F,M,T,\sigma\}$ proceeds as follows:

$S=R^{++}\otimes R^{++}$, $P=R^{+}$,

$F=\{f{:}f\varepsilon F(R,P), \int_0^t \frac{f(\tau)d\tau}{1+(\tau+s)^2}$ exists for every $s,t,\varepsilon R^{++}\}$,

$M=\text{RANGE}(\sigma)$, $T=R^{++}$,

$\sigma(f,t)(x_1,x_2)$

$$=(((1+x_1^2)\times(\exp(2\times\int_0^t \frac{f(\tau)d\tau}{1+(\tau+x_2)^2}))-1)^{1/2}, x_2+t)$$

for every $f\varepsilon F$, $t\varepsilon T$, and $(x_1,x_2)\varepsilon S$. Now condition 2 goes through:

$\sigma(f\to s,t)\sigma(f,s)(x_1,x_2)$

$$=(((1+(1+x_1^2)\times(\exp(2\times\int_0^s \frac{f(\tau)d\tau}{1+(\tau+x_2)^2}))-1)\times$$

$$(\exp(2\times\int_0^t \frac{f\to s(\tau)d\tau}{1+(\tau+x_2+s)^2}))-1)^{1/2},$$

$x_2+s+t)$,

$$=(((1+x_2^2)\times(\exp(2\times\int_0^s \frac{f(\tau)d\tau}{1+(\tau+x_2)^2}))\times$$

$$(\exp(2\times\int_s^{s+t} \frac{f(\tau)d\tau}{1+(\tau+x_2)^2}))-1)^{1/2}, x_2+s+t),$$

$=\sigma(f,s+t)(x_1,x_2)$. Now Z, as redefined, is a system.

The system described by Equation (4) can be thought of as the cascade coupling of two systems Z_1 and Z_2 as illustrated in Figure 5, where $Z_1=\{S_1,P_1,F_1,M_1,T_1,\sigma_1\}$ and $Z_2=\{S_2,P_2,F_2,M_2,T_2,\sigma_2\}$ are defined as follows:

$S_1=R^{++}$,

$P_1=R^{+}\otimes R^{++}$,

$F_1=\{f{:}f\varepsilon\mathscr{Q}(R,P_1), \int_0^t \frac{\pi_1 f(\tau)d\tau}{1+(\pi_2 f(\tau))^2}$ exists for every $t\varepsilon R^{++}\}$,

$M_1=\text{RANGE}(\sigma_1)$,

$T_1=R^{++}$,

$\sigma_1(f,t)(x)=((1+x^2)\times(\exp(2x\int_0^t \frac{\pi_2 f(\tau)d\tau}{1+(\pi_2 f(\tau))^2}))-1)^{1/2}$ for every $f\varepsilon F_1$, $t\varepsilon T_1$ and $x\varepsilon S_1$;

$S_2=R^{++}$,

$P_2=R$,

$F_2=\mathscr{Q}(R,P_2)$

$M_2=\text{RANGE}(\sigma_2)$,

$T_2=R^{++}$,

$\sigma_2(f,t)(x)=x+t$ for every $f\varepsilon F_2$, $t\varepsilon T_2$, $x\varepsilon S_2$

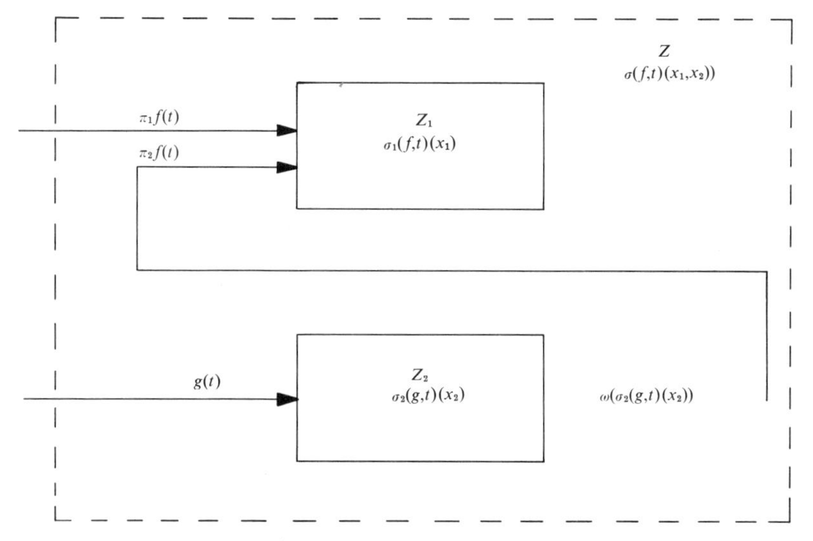

Figure 5

The system Z_2 simply plays the role of an internal clock, where the output function from Z_2 is ω.

These examples should suffice to indicate that both discrete and continuous phenomena can be modeled within the class of models herein developed. Combined discrete and continuous time systems, however, have not been studied at all on the theoretical level nor have the processes of embedding discrete systems in continuous time, discretizing a continuous time system, and approximating a continuous time system by a discrete system been studied as system theoretic manipulations. It is the purpose of the remainder of this paper to begin such studies.

II. THE DISCRETE/CONTINUOUS SYSTEM DICHOTOMY

Discussion

Two approaches to systems theory have been discussed: the approach through finite state machines, automata, Turing machines and the like, and the approach through the theory of differential equations. The first approach has been characterized as the study of discrete time systems, the second as the study of continuous time systems. In the literature to date there is little evidence of these having been considered as two aspects of a single phenomenon. The only clear exception to this statement is the reversion to discrete approximations for the purposes of computing numerical solutions of differential equations.

The postulational approach to systems theory under study in this contribution subsumes both discrete and continuous time systems. When it is said, for example, that $Z=\{S,P,F,M,T,\sigma\}$ is a system, Z could be either discrete or continuous time depending on the way in which T is defined. There is no restraint on T in the postulates except that $T\subset R$ and $0\varepsilon T$.

Even when T is given as I^{++}, for example, there is still no assurance that Z has the classical, or canonical, form of a finite automaton. The essence of the classical discrete system is the recursive definition of $\sigma(f,t)$, that the state of the system at time t depends only on the state and input at time $t-1$. Within the $Z=\{S,P,F,M,T,\sigma\}$ format it is possible to have $T=I^{++}$, but where the state at time t depends on the input experience of z between $t-1$ and t and perhaps on the input experience of Z at times prior to time $t-1$. If $T\subset I^{++}$, it will be said only that Z is time discrete, but if, in addition, Z has the structure of the classical discrete system, it will be said simply that Z is discrete.

The question arises: "In any given modeling or theoretical discussion, to what extent can time-discrete systems be replaced by discrete systems?"

Many phenomena of interest to systems engineers involve interconnections of components, some of which seem to be essentially discrete and some of which seem to be essentially continuous time. An example, and perhaps the archetype, of such systems is the hybrid computer resulting from the interconnection of an analog and a digital computer. If an overall model Z of such a system is desired, T has to be defined reflecting either the discrete or continuous aspect of the system. One way to resolve this dilemma is to define T to reflect the continuous aspect of the system and then to try to embed the model of the discrete components in continuous time.

Now suppose Z is a continuous time system. Quite often it is desirable to derive from Z a discrete system which is more amenable to computation, for example. There are two ways which immediately come to mind for accomplishing this.

One way is to determine a discrete time interval and then simply sample, in a sense, the state of Z at integral values of this time interval. "Under what circumstances does this process, carried out at the theoretical level, yield a time-discrete or discrete system? And in what sense is Z the limit of such sample systems?"

Another way in which to derive a discrete system from a continuous time system is to try to approximate the state of Z by the state of a discrete system Z', rather than to have the state of Z' be the exact state of Z. This approach will have to be placed outside the scope of this paper.

The other questions raised in this introductory discussion will be considered in the theoretical developments in the rest of this paper.

Definition 2.1. Discrete Systems. An assemblage $Z=\{S,P,F,M,T,\sigma\}$ is time discrete if and only if $T\subset I^{++}$.

An assemblage $Z=\{S,P,F,M,T,\sigma\}$ is discrete if and only if Z is time discrete, $1\varepsilon T$, $t\varepsilon T^{+}$ implies $t-1\varepsilon T$, F contains the constant functions, and for every $f\varepsilon F$, $t\varepsilon T$, and $x\varepsilon S$, $\sigma(f,t)(x)$

$=x$ if $t=0$,

$=\sigma(c_{f(t-1)},1)\sigma(f,t-1)(x)$ if $t\varepsilon I^{+}$

Discussion

The first question raised in the introductory discussion involved the relation of a general time-discrete system to the class of discrete systems. There are two ways that immediately come to mind for replacing an arbitrary time-discrete system Z with a discrete system Z'.

One way is simply to feed into Z', as input, the appropriate mapping for Z from M. For example, if Z is started in state x with the input function f, then the state of Z at time $t\varepsilon I^{++}$ is $\sigma(f,t)(x)$. Let Z' be defined as a discrete system with the same set of states as Z but such that the inputs to Z' are the elements of M, the set of mappings of S into itself which determines the behavior of Z, $P'=M$. Let F' be the admissible set of input functions generated by the constant functions with values in P'. Now Z' will exhibit the same behavior as Z if, given $f\varepsilon F$, the input g is introduced into Z', where g is defined as follows: g is a step function and for every $t\varepsilon I$, $g(t)$

$=\omega$ if $t\varepsilon I^{-}$,

$=\sigma(f\rightarrow t,1)$ if $t\varepsilon I^{++}$. In other words, the information which Z' needs in order to ape Z's behavior at time t is fed to Z' as input at time $t-1$.

Theoretically this procedure is legitimate, but the practical difficulty with this procedure is that the real computation of the relation between the behavior of Z and its input is left to the experimenter who generates the input function g.

Another approach to this problem, one that overcomes the objection just raised to the definition of Z', is as follows: instead of feeding the correct mappings for Z to Z', feed the input for Z to a system Z'' on a stepped-up time scale and let Z'' compute its own appropriate mapping. This process can be understood as follows: suppose the state of Z at time t depends on all of f between time $t-1$ and t; then, if Z'' is supposed to take on the same state at time t as Z does, Z'' will have to have all the information about f between $t-1$ and t *at time* $t-1$. So, instead of feeding f continuously to Z'', so to speak, f is fed to Z'' in dollops as a step function. At time 0, Z'' should receive as its input the whole segment of f between 0 and 1; the input to Z'' can then be constant between 0 and 1 but at time 1, Z'' must receive all the segment of f between 1 and 2 and so on. Then, at time 0, Z'' has everything it needs to compute Z's state for time 1 and so on. What this all means, of course, is that $P''=\{p\colon p\varepsilon\mathcal{Q}(R[0,1),P)$, there exists $f\varepsilon F$ such that $p=\text{res}(f,R[0,1))\}$ and F'' is the set of all step functions with countably many steps.

Theorem 2.1. Replacing Time-Discrete Systems by Discrete Systems. Let Z be a time-discrete system such that $T=I^{++}$.

Let $Z'=\{S',P',F',M',T',\sigma'\}$ be specified as follows: $S'=S$, $P'=M$, F' is the set of all functions on R with values in P' which are step functions with a finite or countable number of steps, $M'=\text{RANGE}(\sigma')$, $T'=T$ and for every $f\varepsilon F'$, $t\varepsilon T'$, $x\varepsilon S'$, let

$\sigma'(f,t)(x)$

$=x$ if $t=0$,

$=\sigma'(c_{f(t-1)},1)(\sigma'(f,t-1)(x))$ if $t\varepsilon I^+$,

and for every $p\varepsilon P'$, let $\sigma'(c_p,1)=p$.

Then Z' is a discrete system and if $f\varepsilon F$ and $x\varepsilon S$ are given, let g be any function in F' such that $g(t)=\sigma(f\rightarrow t,1)$ for every $t\varepsilon T$; then $\sigma'(g,t)(x)$ $=\sigma(f,t)(x)$ for every $t\varepsilon T$.

Let Z'' be defined as follows: $S''=S$, $P''=\{p\colon p\varepsilon\mathcal{Q}(R[0,1),P)$, there exists $f\varepsilon F$ such that $p=\text{res}(f,R[0,1))\}$, F'' is the set of functions defined on R with values in P'' which are step functions with a finite or countable number of steps, $M''=\text{RANGE}(\sigma'')$, $T''=T$, and for every $f\varepsilon F''$, $t\varepsilon T''$, $x\varepsilon S''$, let $\sigma''(f,t)(x)$

$=x$ if $t=0$,

$=\sigma''(c_{f(t-1)},1)(\sigma''f,t-1)(x))$ if $t\varepsilon I^+$ and for every $p\varepsilon p''$, let $\sigma''(c_p,1)$ $=\sigma(f,1)$ where $p=\text{res}(f,R[0,1))$.

Then Z'' is a discrete system and if $f \varepsilon F$ and $x \varepsilon S$ are given, let h be any function in F'' such that $h(t) = \text{res}(f \rightarrow t, R[0,1))$ for every $t \varepsilon T$; then $\sigma''(g,t)(x) = \sigma(f,t)(x)$ for every $t \varepsilon T$.

Proof: It isn't hard to see that both F' and F'' are admissible sets of input functions with values in P' and P'', respectively, and with this observation, it follows that Z' and Z'' are discrete systems.

Now assume $f \varepsilon F$ and $x \varepsilon S$ are given. Let $g \varepsilon F'$, $g(t) = \sigma(f \rightarrow t, 1)$ for every $t \varepsilon T$. Then prove that $\sigma'(g,t)(x) = \sigma(f,t)(x)$ for every $t \varepsilon T$ by induction:

$\sigma'(g,0)(x) = x = \sigma(f,0)(x)$. Assume that $n \varepsilon I^+$ and that $\sigma'(g,t)(x) = \sigma(f,t)(x)$ for every $t \varepsilon I[0,n]$.

Then $\sigma'(g,n+1)(x)$
$= \sigma'(c_{g(n)},1)(\sigma'(g,n)(x))$ (by the recursive definition of σ'),
$= g(n)(\sigma'(g,n)(x)))$ by the definition of $\sigma'(c_p,1)$),
$= \sigma(f \rightarrow n,1)(\sigma(f,n)(x))$ (by the definition of g and the induction hypothesis),
$= \sigma(f,n+1)(x)$ (because Z is a system).

Now show by induction that
$\sigma''(h,t)(x) = \sigma(f,t)(x)$ where $h \varepsilon F''$ and $h(t) = \text{res}(f \rightarrow t, R[0,1))$ for every $t \varepsilon T$:
$\sigma''(h,0)(x) = x = \sigma(f,0)(x)$. Assume that $n \varepsilon I^+$ and that $\sigma''(h,t)(x) = \sigma(f,t)(x)$ for every $t \varepsilon I[0,n]$. Then
$\sigma''(h,n+1)(x)$
$= \sigma''(c_{h(n)},1)(\sigma''(h,n)(x))$ (by the recursive definition of σ''),
$= \sigma''(c_{\text{res}(f \rightarrow n, R[0,1))},1)(\sigma''(h,n)(x))$ (by the definition of h),
$= \sigma(f \rightarrow n,1)(\sigma(f,n)(x))$ (by the definition of $\sigma''(c_p,1)$ and the induction hypothesis),
$= \sigma(f,n+1)(x)$ (because Z is a system).

Discussion

Ideally, Theorem 2.1 should produce a system Z''' such that given $f \varepsilon F$ and $x \varepsilon S$, then $f \varepsilon F'''$ and $\sigma'''(f,t)(x) = \sigma(f,t)(x)$ for every $t \varepsilon T$. The Theorem 2.1 which has just been proved falls a little short of this ideal. In the cases of both Z' and Z'', some preprocessing of f has to take place in order that Z' and Z'' are able to ape the behavior of Z. But the ideal is impossible of attainment, as the following example shows.

Let Z be defined as follows: $S = R$, $P = R$, F is the admissible set of input functions generated by the continuous functions with values in P, $M = \text{RANGE}(\sigma)$, $T = I^{++}$, $\sigma(f,t)(x) = x + \int_0^t f(\tau)d\tau$. Then Z is a time-discrete system because $T = I^{++}$, but Z is not discrete. Suppose there exists Z''' such that $S''' = S$, $P''' = P$, $F''' = F$, $M''' = \text{RANGE}(\sigma''')$, $T''' = I^{++}$,

$\sigma'''(f,0)=\omega$, $\sigma'''(f,t)(x)=\sigma'''(c_{f(t-1)},1)\sigma'''(f,t-1)(x)$ for every $f \varepsilon F'''$, $t \varepsilon I^+$, $x \varepsilon S'''$, and such that for every $f \varepsilon F$, $s \varepsilon S$, $\sigma'''(f,t)(x)=\sigma(f,t)(x)$ for every $t \varepsilon T$. To be specific, let $x=1$, $f_1(t)=t$ and $f_2(t)=-t$ for every $t \varepsilon R$. Then

$\sigma(f_1,1)(x)$

$=1+\int_0^1 f_1(\tau)d$ (because $x=1$),

$=\frac{3}{2}$ (because $f_1(t)=t$); and

$\sigma(f_2,1)(x)=\frac{1}{2}$. But

$\sigma'''(f_1,1)(x)$

$=\sigma'''(c_{f_1(0)},1)(x)$ (by the recursive definition of σ'''),

$=\sigma'''(c_0,1)(x)$ (by the definition of f_1),

$=\sigma'''(c_{f_2(0)},1)(x)$ (by the definition of f_2),

$=\sigma'''(f_2,1)(x)$ (by the definition of σ''')

Hence $\sigma'''(c_0,1)(x)$ cannot be both $\sigma(f_1,1)(x)$ and $\sigma(f_2,1)(x)$. Therefore the ideal Z''' does not exist.

But the constructions of Theorem 2.1 can be useful in any modeling or simulation situation to indicate, at least, how to restrict the discussion of time-discrete phenomena to discrete models. But such examples, however, motivate exploration of the possibility of discrete systems approximating given systems.

The next question which will be taken up is that of embedding a discrete system in continuous time. This manipulation is important in modeling phenomena which are interconnections of components, some of which are easier to model independently as continuous time systems, some of which are easier to model as discrete systems.

Suppose given a discrete system Z and attempt to construct a system Z' such that $T'=R^{++}$ and such that for every $f \varepsilon F$ and $x \varepsilon S$ there exists a $g \varepsilon F'$ and a $y \varepsilon S'$ such that $\sigma(f,n)(x)=\sigma'(g,n)(y)$ for every $n \varepsilon T$.

It isn't possible to do quite that well with Z' because the states of Z' have to be more complicated in consisting of more components than just the states of Z. For one thing, if the state of Z' at time $n+1$ depends only on its input at time n, then, since Z' exists in continuous time, Z' has to remember the input it received at time n during the interval of time from n to $n+1$. Another component of the state variable of Z' has essentially to be a clock to time continuously the discrete behavior of Z'. The requirements of the input memory and the clock component of the state of Z' do not arise merely to satisfy intuition founded on the hardware (analog) implementation of discrete systems but apparently must also be present in order that Z' satisfy the system requirements of Definition 1.1.

Theorem 2.2 Embedding a Discrete System in Continuous Time. Let Z be a discrete system such that $T=I^{++}$ and F is the admissible set of input functions with values in P generated by the constant functions.

Define Z' as follows: $S'=S\otimes P\otimes R^{++}$, $P'=P$, $F'=F$, $M'=\text{RANGE}(\sigma')$, $T'=R^{++}$, and, for every $f \in F'$, $(x,p,u) \in S'$, $t \in T'$, define $\sigma'(f,t)(x,p,u)$ as follows:

$\sigma'(f,t)(x,p,u)$
$=(x,p,u+t)$ if $t+\text{FRACTION}(u) \in R(1)$,
$=(\sigma(\text{step}(c_p,1,f\rightarrow -\text{FRACTION}(u)), \text{INTEGER}(u+t)-\text{INTEGER}(u))(x),$
$f((\text{INTEGER}(u+t)-u)^-),u+t)$ if $t+\text{FRACTION}(u)\geq 1$,

where, for every $s \in R^{++}$, $\text{INTEGER}(s)$ and $\text{FRACTION}(s)$ are uniquely determined as follows: $\text{INTEGER}(s) \in I^{++}$, $\text{FRACTION}(s) \in R(1)$, $s=\text{INTEGER}(s)+\text{FRACTION}(s)$, and $f(s^-)$ is the limit of f from the left at s.

Then Z' is a system and, for every $f \in F$, $x \in S$, $\sigma(f,t)(x)=\pi_1(\sigma'(f,t)(x,f(0),0))$ for every $t \in T$.

Proof. It's clear that Z' is an assemblage. Furthermore, it's clear that $\sigma'(f,0)=\omega$ for every $f \in F'$, since, if $t=0$, $t+\text{FRACTION}(u) \in R(1)$.

Now let $f \in F'$, $s,t \in T$, $(x,p,u) \in S'$, then it must be shown that $\sigma'(f\rightarrow s,t)(\sigma'(f,s)(x,p,u))$
$=\sigma'(f,s+t)(x,p,u)$. In the computation which follows the fact is used that, for Z, if $f, g \in F$ and $f(s)=g(s)$ for every $s \in I[0,t)$, then $\sigma(f,t)=\sigma(g,t)$. Now consider:

$\sigma'(f\rightarrow s,t)(\sigma'(f,s)(x,p,u))$
$=\sigma'(f\rightarrow s,t)(x,p,u+s)$ if $s+\text{FRACTION}(u) \in R(1)$,
$=\sigma'(f\rightarrow s,t)(\sigma(\text{step}(c_p,1,f\rightarrow -\text{FRACTION}(u)),\text{INTEGER}(u+s)-\text{INTEGER}(u)).$
$(x),f((\text{INTEGER}(u+s)-u)^-),u+s)$ if $s+\text{FRACTION}(u)\geq 1$;
$=(x,p,u+s+t)$ if $s+\text{FRACTION}(u) \in R(1)$, $t+\text{FRACTION}(u+s) \in R(1)$,
$=(\sigma(\text{step}(c_p,1,f\rightarrow s\rightarrow -\text{FRACTION}(u+s)), \text{INTEGER}(u+s+t)-$
$\text{INTEGER}(u+s))(x), f\rightarrow s((\text{INTEGER}(u+s+t)-u-s)^-),u+s+t)$ if
$s+\text{FRACTION}(u) \in R(1)$, $t+\text{FRACTION}(u+s)\geq 1$,
$=(\sigma(\text{step}(c_p,1,\rightarrow -\text{FRACTION}(u)), \text{INTEGER}(u+s)-\text{INTEGER}(u))(x),$
$f((\text{INTEGER}(u+s)-u)^-),u+s+t)$ if $s+\text{FRACTION}(u)\geq 1$,
$t+\text{FRACTION}(u+s) \in R(1)$,
$=(\sigma(\text{step}(c_{f((\text{INTEGER}(u+s)-u)^-)},1,f\rightarrow s\rightarrow -\text{FRACTION}(u+s)),$
$\text{INTEGER}(u+s+t)-\text{INTEGER}(u+s))(\sigma(\text{step}(c_p,1,f\rightarrow -\text{FRACTION}(u)),$
$\text{INTEGER}(u+s)-\text{INTEGER}(u)),(x)),f\rightarrow s((\text{INTEGER}(u+s+t)-u-s)^-),$
$u+s+t)$ if $s+\text{FRACTION}(u)\geq 1$, $t+\text{FRACTION}(u+s)\geq 1$;
$=(x,p,u+s+t)$ if $s+\text{FRACTION}(u) \in R(1)$, $t+\text{FRACTION}(u+s) \in R(1)$,

$=(\sigma(\text{step}(c_p,1,f\to -\text{FRACTION}(u)), \text{INTEGER}(u+s+t)-\text{INTEGER}(u))(x),$
$f((\text{INTEGER}(u+s+t)-u)^-), u+s+t)$
$\text{if } s+\text{FRACTION}(u)\in R(1),\ t+\text{FRACTION}(u+s)\geq 1$
(because, if $s+\text{FRACTION}(u)\in R(1)$, then
$\text{INTEGER}(u+s)=$
$\text{INTEGER}(u)$ and
$\text{FRACTION}(u+s)=s+\text{FRACTION}(u)$ and
$f\to s(r^-)=f((s+r)^-)$ for every $r\in R$),
$=(\sigma(\text{step}(c_p,1,f\to -\text{FRACTION}(u)), \text{INTEGER}(u+s+t)-\text{INTEGER}(u))(x),$
$f((\text{INTEGER}(u+s+t)-u)^-),u+s+t)$ if $s+\text{FRACTION}(u)\geq 1,$
$t+\text{FRACTION}(u+s)\in R(1)$ (because, if
$t+\text{FRACTION}(u+s)\in R(1)$, then
$\text{INTEGER}(u+s)=\text{INTEGER}(u+s+t)$),
$=(\sigma(((\text{step}(c_p,1,f\to -\text{FRACTION}(u))\to(\text{INTEGER}(u+s)-\text{INTEGER}(u)))|$
$(\text{step}(c_{f((\text{INTEGER}(u+s)-u)^-)},1,f\to s\to -\text{FRACTION}(u+s))))$
$\to(\text{INTEGER}(u)-\text{INTEGER}(u+s)), \text{INTEGER}(u+s+t)-\text{INTEGER}(u))(x),$
$f((\text{INTEGER}(u+s+t)-u)^-), u+s+t)$ if $s+\text{FRACTION}(u)\geq 1,$
$t+\text{FRACTION}(u+s)\geq 1$ (because, for any g, $h\in F$, q, $r\in T$,
$\sigma(h,q)\sigma(g,r)$
$=\sigma(((g\to r)\ |h)\to -r,r+q))$;
$=(x,p,u+s+t)$ if $s+t+\text{FRACTION}(u)\in R(1),$
$=(\sigma(\text{step}(c_p,1,f\to -\text{FRACTION}(u)), \text{INTEGER}(u+s+t)-\text{INTEGER}(u))(x),$
$f((\text{INTEGER}(u+s+t)-u)^-), u+s+t)$ if $s+\text{FRACTION}(u)\in R(1),$
$s+t+\text{FRACTION}(u)\geq 1$ or if $s+\text{FRACTION}(u)\geq 1,$
$t+\text{FRACTION}(u+s)\in R(1),$
$=(\sigma(\text{step}(c_p,1,f\to -\text{FRACTION}(u)), \text{INTEGER}(u+s+t)-\text{INTEGER}(u))(x),$
$f((\text{INTEGER}(u+s+t)-u)^-), u+s+t)$ if $s+\text{FRACTION}(u)\geq 1,$
$t+\text{FRACTION}(u+s)\geq 1$ (because, for every $r\in I$,
$(((\text{step}(c_p,1,f\to -\text{FRACTION}(u))\to(\text{INTEGER}(u+s)-$
$\text{INTEGER}(u)))|\text{step}(c_{f((\text{INTEGER}(u+s)-u)^-)},1,f\to s\to -$
$\text{FRACTION}(u+s)))\to(\text{INTEGER}(u)-\text{INTEGER}(u+s)))(r)$
$=(\text{step}(c_p,1,f\to -\text{FRACTION}(u)))(r))$;
$=\sigma'(f,s+t)(x,p,u)$ in every case.

Now assume that $t\in T'$ and $f,g\in F'$ such that $\text{res}(f,R(t))$ $=\text{res}(g,R(t))$. Then for $(x,p,u)\in S'$:

$\sigma'(f,t)(x,p,u)$
$=(x,p,u+t)$ if $t+\text{FRACTION}(u)\in R(1),$

$=(\sigma(\text{step}(c_p,1,f\rightarrow-\text{FRACTION}(u)),\ \text{INTEGER}(u+t)-\text{INTEGER}(u))(x),$
$f((\text{INTEGER}(u+t)-u)^-),\ u+t)$ if $t+\text{INTEGER}(u)\geq 1$;
$=(x,p,u+t)$ if $t+\text{FRACTION}(u)\varepsilon R(1)$,
$=(\sigma(\text{step}(c_p,1,g\rightarrow-\text{FRACTION}(u)),\text{INTEGER}(u+t)-\text{INTEGER}(u))(x),$
$g((\text{INTEGER}(u+t)-u)^-),\ u+t)$ if $t+\text{INTEGER}(u)\geq 1$
(because, for every $s\,\varepsilon\,R(t)$,
$(\text{step}(c_p,1,f\rightarrow-\text{FRACTION}(u))(s)=(\text{step}(c_p,1,g\rightarrow-\text{FRACTION}(u)))(s)$, and $\text{INTEGER}(u+t)-u\,\varepsilon\,R[0,t]$, hence
$f((\text{INTEGER}(u+t)-u)^-)=g((\text{INTEGER}(u+t)-u)^-))$,
$=\sigma'(g,t)(x,p,u)$. Hence Z' is a system.

Now let $f\,\varepsilon\,F$, $x\,\varepsilon\,S$, $t\,\varepsilon\,T$, then:
$\pi_1(\sigma'(f,t)(x,f(0),0))$
$=\pi_1(x,f(0),0)$ if $t=0$,
$=\pi_1(\sigma(\text{step}(c_{f(0)},1,f\rightarrow-\text{FRACTION}(0)),\text{INTEGER}(0+t)-\text{INTEGER}(0))(x),$
$f((\text{INTEGER}(0+t)-0)^-),0+t)$ if $t\,\varepsilon\,I^+$;
$=x$ if $t=0$,
$=\sigma(f,t)(x)$ if $t\,\varepsilon\,I^+$;
$=\sigma(f,t)(x)$ for all $t\,\varepsilon\,T$.

Discussion

The embedding of a discrete system in continuous time, given by Theorem 2.2, is achieved by brute force, so to speak. The state space is still finite and/or topologically discrete except for the time component itself. For many purposes this may be satisfactory and, in fact, might be better for some purposes than embedding the discrete system in a system whose state space is a continuum in every component. On the other hand, it would be useful to be able to embed a discrete Z in a continuous time Z' whose time trajectories are defined by differential equations. At this moment, however, that possibility represents a difficult problem requiring more mathematical machinery than is presently available.

For computational purposes, it is often desirable to replace a continuous time system by a time-discrete system. Even on the theoretical level, this can be accomplished by sampling the state of the original system at discrete time intervals in order to define the time trajectories of a time-discrete system. This is a way, incidentally, in which systems may arise naturally which are time discrete but not discrete.

Theorem 2.3. Sampling a Continuous Time System to Generate a Time-Discrete System. Let Z be a system such that $T=R^{++}$. For each $d\,\varepsilon\,R^+$, let Z_d be defined as follows: $S_d=S$, $P_d=P$, $F_d=\{g: g\,\varepsilon\,\mathcal{Q}(R,P),$

there exists $f \varepsilon F$ such that $g(t)=f(t \times d)$ for every $t \varepsilon R\}$, $M=\text{RANGE}(\sigma_d)$, $T_d=I^{++}$, for every $g \varepsilon F_d$, $x \varepsilon S_d$, $n \varepsilon T_d, \sigma_d(g,n)(x)=\sigma(f,n \times d)(x)$, where $f \varepsilon F$ and $g(t)=f(t \times d)$ for every $t \varepsilon R$.

Then Z_d is a system.

Let $\mu_d \varepsilon \mathcal{Q}(F,F_d)$ be defined as follows for $f \varepsilon F$: $(\mu_d(f))(t)=f(t \times d)$ for every $t \varepsilon R$.

Then μ_d is one to one and onto and for every $f,g \varepsilon F$ and $r \varepsilon R$, $\mu_d(f \rightarrow r)=\mu_d(f) \rightarrow \frac{r}{d}$, $\mu_d(f|g)=\mu_d(f)|\mu_d(g)$ and $\sigma(f,n \times d)(x)=\sigma_d(\mu_d(f),n)(x)$ for every $n \varepsilon I^{++}$.

Given $f \varepsilon F$ and $x \varepsilon S$, let $\eta \varepsilon \mathcal{Q}(R^+ \otimes I^{++},S)$ be defined as follows for $(d,n) \varepsilon R^+ \otimes I^{++}$:

$\eta(d,n)=\sigma_d(\mu_d(f),n)(x)$. Let $\alpha \varepsilon \mathcal{Q}(\text{DOMAIN}(\eta)^2, \{0,1\})$ be defined as follows for $((d,m),(e,n)) \varepsilon \text{DOMAIN}(\eta)^2$:

$\alpha((d,m),(e,n))$

$=1$ if $e<d$, and $n>m$,

$=0$ otherwise. Then α is a partial order on $\text{DOMAIN}(\eta)$ and $\text{DOMAIN}(\eta)$ is directed by α.

Then $\{\eta,\alpha\}$ is a net in S and every state in the time trajectory of Z determined by $f \varepsilon F$ and $x \varepsilon S$ is a cluster point of $\{\eta,\alpha\}$ and for every $t \varepsilon T$, there exists a subnet of $\{\eta,\alpha\}$ which converges to $\sigma(f,t)(x)$ regardless of the topology on S.

Proof: To show that Z_d is a system it must first be shown that F_d is an admissible set of input functions with values in P_d. Let $f,g \varepsilon F_d$, $r \varepsilon R$, then there exists $f',g' \varepsilon F$ such that $f(t)=f'(t \times d)$ and $g(t)=g'(t \times d)$ for every $t \varepsilon R$. Then

$(f \rightarrow r)(t)$

$=f(r+t)$,

$=f'((r+t) \times d)$ (by the choice of f'),

$=(f' \rightarrow r \times d)(t \times d)$,

which shows that $f \rightarrow r \varepsilon F_d$ because $f' \rightarrow r \times d \varepsilon F$ if $f' \varepsilon F$. Similarly,

$(f|g)(t)$

$=f(t)$ if $t \varepsilon R^-$,

$=g(t)$ if $t \varepsilon R^{++}$;

$=f'(t \times d)$ if $t \varepsilon R^-$,

$=g'(t \times d)$ if $t \varepsilon R^{++}$;

$=(f'|g')(t \times d)$,

which shows that $f|g \varepsilon F_d$ because $f'|g' \varepsilon F$ if f' and $g' \varepsilon F$. Hence Z_d is an assemblage.

To show that Z_d is a system, let $f,g \varepsilon F_d$ and $f',g' \varepsilon F$ such that $f(t)=f'(t \times d)$, $g(t)=g'(f \times d)$ for every $t \varepsilon R$; let $s \varepsilon R$, then

$\sigma_d(f,0)=\sigma(f',0)=\omega$ because Z is a system.

Also, if $s,t,s+t \varepsilon T_d$:

$\sigma_d(f \rightarrow s,t)\sigma_d(f,s)$

$=\sigma(f' \rightarrow s \times d, t \times d)\sigma(f',s \times d)$ (by the definition of σ_d, f' and the above deduction where it is shown that $(f \rightarrow s)(u)=(f' \rightarrow s \times d)(u \times d)$ for every $y \varepsilon R$),

$=\sigma(f', t \times d+s \times d)$ (because Z is a system and $t \times d$, $s \times d$, $t \times d+s \times d \varepsilon T$),

$=\sigma(f', (s+t) \times d)$,

$=\sigma_d(f,s+t)$ (by the definition of σ_d).

Now assume that $t \varepsilon T_d$, res$(f,R(t))=$res$(g,R(t))$.

Then

$\sigma_d(f,t)$

$=\sigma(f', t \times d)$ (by the definition of σ_d and the choice of f'),

$=\sigma(g', t \times d)$ (because Z is a system and if res$(f,R(t))=$res$(g,R(t))$ then res$(f',R(t \times d))=$res$(g',R(t \times d))$),

$=\sigma_d(g,t)$ (by the definition of σ_d and the choice of g').

Hence Z_d is a system.

By the definition of F_d, μ_d is onto. If $f,g \varepsilon F$ and $t \varepsilon R$ such that $f(t) \neq g(t)$, then

$(\mu(f))\left(\frac{t}{d}\right)$

$=f(t)$ (by the definition of μ_d),

$\neq g(t)$ (by assumption),

$=(\mu(g))\left(\frac{t}{d}\right)$;

hence μ_d is one to one. Now assume $f,g \varepsilon F$ and $r \varepsilon R$, then:

$(\mu_d(f \rightarrow r))(t)$

$=(f \rightarrow r)(t \times d)$,

$=f(r+t \times d)$,

$=f(\left(\frac{r}{d}+t\right) \times d)$,

$=(\mu_d(f))\left(\frac{r}{d}+t\right)$,

$=\left(\mu_d(f) \rightarrow \frac{r}{d}\right)(t)$;

$(\mu_d(f|g))(t)$

$=f(t \times d)$ if $t \times d \varepsilon R^-$,

$=g(t \times d)$ if $t \times d \varepsilon R^{++}$;

$=(\mu_d(f))(t)$ if $t \times d \varepsilon R^-$,

$=(\mu_d(g))(t)$ if $t \times d \varepsilon R^{++}$;

$=(\mu_d(f)|\mu_d(g))(t)$ (because $d \varepsilon R^+$).

Therefore, $\mu_d(f \to r) = \mu_d(f) \to \frac{r}{d}$ and $\mu_d(f|g)$ $= \mu_d(f)|\mu_d(g)$ as asserted.

If $f \varepsilon F$ and $n \varepsilon I^{++}$, then $(\mu_d(f))(t) = f(t \times d)$ for every $t \varepsilon R$, therefore, by definition of σ_d, $\sigma_d(\mu_d(f),n) = \sigma(f,n \times d)$.

To show that α is a partial order on DOMAIN(η), it must be shown that if $(d,m),(d',m'),(d'',m'') \varepsilon$ DOMAIN(η), then
$\alpha((d,m),(d,m)) = 1$ and if
$\alpha((d,m),(d',m')) = \alpha((d',m'),(d'',m'')) = 1$, then
$\alpha((d,m),(d'',m'')) = 1$. It's clear that
$\alpha((d,m),(d,m)) = 1$. If
$\alpha((d,m),(d',m')) = \alpha((d',m'),(d'',m'')) = 1$, then,
by the definition of α,
$d' \leq d$, $m' \geq m$, $d'' \leq d'$, $m'' \geq m'$ so that $d'' \leq d$ and $m'' \geq m$ and hence $\alpha((d,m),(d'',m'')) = 1$' Hence α is a partial order. Furthermore if (d,m), $(d',m') \varepsilon$ DOMAIN(η) are arbitrary and $e = \wedge\{d,d'\}$, $n = \vee\{m,m'\}$, then $\alpha((d,m),(e,n)) = \alpha((d',m'),(e,n)) = 1$, and hence DOMAIN($\eta$) is directed by α. Therefore, $\{\eta,\alpha\}$ is a net in S.

Now let $t \varepsilon T$ and show that $\sigma(f,t)(x)$ is a cluster point of $\{\eta,\alpha\}$. Formally, it must be shown that if $N \varepsilon$ NEIGHBORHOODS ($\sigma(f,t)(x)$, TOPOLOGY(S)), and $(d,m) \varepsilon$ DOMAIN(η), then there exists $(e,n) \varepsilon$ DOMAIN(η) such that $\alpha((d,m),(e,n)) = 1$ and $\eta(e,n) \varepsilon N$. To this end, given $(d,m) \varepsilon$ DOMAIN(η), let n
$= m$ if $d \times m \geq t$,

$= \wedge\{p: p \varepsilon I^+, p \geq \frac{t}{d}\}$ if $d \times m$

$< t$, let $e = \frac{t}{n}$. Then $t = e \times n \varepsilon T$ and hence $n \varepsilon T_e$, $e_e R^+$ and e

$= \frac{t}{n}$ (by definition);

$= \frac{t}{m}$ if $d \times m \geq t$,

$= \dfrac{t}{\wedge\{p: p \varepsilon I^+, p \geq \frac{t}{d}\}}$ if $d \times m < t$ (by definition of n);

$\leq d$ if $d \times m \geq t$,

$\leq \dfrac{t}{\frac{t}{d}}$ if $d \times m < t$; and, further, since n

$= m$ if $d \times m \geq t$,

$= \wedge\{p: p \varepsilon I, p \geq \frac{t}{d}\}$ if $d \times m < t$;

$n \geq m$ in all cases; hence $\alpha((d,m),(e,n)) = 1$.

Furthermore,
$\eta(e,n)$
$=\sigma_e(\mu_e(f),n)(x)$ (by the definition of η),
$=\sigma(f,n\times e)(x)$ (by the definition of σ_e),
$=\sigma(f,t)(x)$ (by the definition of e).
Therefore, $\eta(e,n)$ is not only in N but actually equal to $\sigma(f,t)(x)$.

To construct a subnet which converges to $\sigma(f,t)(x)$, let $B_t=\{(d,m)$: $(d,m)\varepsilon$DOMAIN(η), $d\times m=t\}$ directed by $\beta_t=\text{res}(\alpha,B_t)$; let $\theta_t=\text{res}(\eta,B_t)$. Then $\{\theta_t,B_t\}$ is a subnet of $\{\eta,\alpha\}$: let γ_t be the function defined on B_t with values in DOMAIN(η) as follows for $(d,m)\varepsilon B_t$: $\gamma_t(d,m)=(d,m)$. Then
$\theta_t(d,m)$
$=\eta(d,m)$,
$=\eta\gamma(d,m)$, hence $\theta_t=\eta\gamma_t$; given $(d,m)\varepsilon$DOMAIN(η), there must exist $(e,n)\varepsilon B_t$ such that if $\beta_t((e,n),(e',n'))=1$ then $\alpha((d,m),\gamma(e',n'))=1$; the (e,n) defined in terms of (d,m) does the trick. It is clear that for all $(d,m)\varepsilon B_t$, $\theta_t(d,m)=\sigma(f,t)(x)$.

Discussion

In terms of the Z given in Theorem 2.3, define Z_d' as follows: $S_d'=S$, $P_d'=P$, $F_d'=F$, $M_d'=$RANGE(σ_d'), $T_d'=\{n\times d$: $n\varepsilon I^{++}\}$, $\sigma_d'(f,t)=\sigma(f,t)$ for every $f\varepsilon F_d'$, $t\varepsilon T_d'$. Then Z_d' is a subsystem of Z. The point, then, of the second paragraph of assertions in Theorem 2.3 is to assert that Z_d is isomorphic $(1,\mu_d,\omega)$ to the subsystem Z_d' of Z. And in this sense, Z simulates all the Z_d's.

The last set of assertions in Theorem 2.3 wants to say, formally, that the sampling process produces a net in the state space of Z which, for each $t\varepsilon T$, contains a subnet which converges to $\sigma(f,t)(x)$. In this particular case, the statement is an example of a nontheorem because it's clear that convergence, in this case, is perfectly trivial. But the wording and format are appropriate for theorems of this kind. In the case of a true approximation theorem this format will be appropriate and nontrivial.

Further approximation theorems will have to await further advances in general systems theory.

APPENDIX

Specialized Notation

If A and B are sets, then $\mathscr{F}(A,B)$ is the set of all functions defined on A with values in B.

If $f \varepsilon \mathscr{F}(A,B)$ and $C \subset A$, then res(f,C) denotes the restriction of f to C; the restriction of f to C is a function defined on C with values in B defined as follows for every $a \varepsilon C$: $(\text{res}(f,C))(a)=f(a)$.

If $f \varepsilon \mathscr{F}(A,B)$ then $\Delta(f)$ and $\Gamma(f)$ denote, respectively, the domain and range of f and are defined as follows: $\Delta(f)=A$, $\Gamma(f)=\{y:y \varepsilon B$, there exists $x \varepsilon A$ such that $y=f(x)\}$.

The set of all real numbers is denoted R. Subsets of R are denoted and defined as follows: If $r,s \varepsilon R$,

$R(r,s)$
$=\{t:t \varepsilon R,\ r<t\leq s\}$ if $r<s$,
$=\Phi$ otherwise;
$R(r,s]$
$=\{t:t \varepsilon R, r<t\leq s\}$ if $r<s$,
$=\Phi$ otherwise;
$R[r,s)$
$=\{t:t \varepsilon R. r\leq t<s\}$ if $r<s$,
$=\Phi$ otherwise;
$R[r,s]$
$=\{t:t \varepsilon R, r\leq t\leq s\}$ if $r<s$,
$=\Phi$ otherwise;
$R(s)$
$=R[0,s)$ if $s>0$,
$=\{0\}$ if $s=0$
$=R[s,0)$ if $s<0$;
$R^+=\{t:t \varepsilon R,\ t>0\}$;
$R^-=\{t:t \varepsilon R,\ t<0\}$;
$R^{++}=R^+ \cup \{0\}$.

If T is a subset of R, then $T(r,s)=T \cap R(r,s)$, and so on, and $T^+=T \cap R^+$, for example. The set of integers, positive, negative, and 0 is denoted I.

If P is a set and $p \varepsilon P$, then c_p is a function defined on R with values in P as follows for every $t \varepsilon R$: $c_p(t)=p$.

If P is a set, $n \varepsilon I^+$, $\{f_o,\cdots,f_n\} \subset \mathscr{F}(R,P)$, $\{t_1,\cdots,t_n\} \subset R$ such that $t_1<\cdots<t_n$, then step$(f_o,t_1,\cdots,t_n, f_n)$ denotes a function defined on R with values in P defined as follows for every $t \varepsilon R$:

$(\text{step}(f_o,t_1,\cdots,t_n,f_n))(t)$
$=f_o(t)$ if $t<t_1$,

$= f_i(t)$ if $t \in R(t_i, t_{i+1})$, $i \in I[1, n-1]$,
$= f_n(t)$ if $t \geq t_n$.

If A is a set then $\omega \in \mathcal{Q}(A, A)$ and is defined as follows for every $a \in A$: $\omega(a) = a$.

If $A_1, \cdots, A_n$ are sets, then $A_1 \otimes \cdots \otimes A_n$ denotes the set of all n-tuples $(a_1, \cdots, a_n)$ where $a_i \in A_i$ for every $i \in I[1, n]$. For each $i \in I[1, n]$, π_i denotes a function defined on $A_1 \otimes \cdots \otimes A_n$ with values in A_i as follows for $(a_1, \cdots, a_n) \in A_1 \otimes \cdots \otimes A_n$: $\pi_i(a_1, \cdots, a_n) = a_i$.

If A is a set and $n \in I^+$ then A_n denotes the set of all n-tuples $(a_1, \cdots, a_n)$ where $a_i \in A$ for every $i \in I[1, n]$.

2

MATHEMATICAL THEORY OF GENERAL SYSTEMS

MIHAJLO D. MESAROVIĆ

I. INTRODUCTION

In the last two decades a new branch of applied mathematics dealing with the problems of information processing (computers) and decision making (control) has been under very rapid development. In general, this development was motivated by the problems of automation in a broad sense (in industry, business, and administration) i.e., with the substitution by machinery (i.e., computers) of the activities traditionally in the domain of human intelligence. To be able to use machines effectively one has to rely to a lesser degree on intuitive understanding and will have to make explicit many tacitly assumed conditions and circumstances. This, of course, requires the development of a formal framework.

In spite of a large diversity of problems and applications, there are some concepts which are common in almost all areas of the information-processing and decision-making field. Such a central concept is that of a system. In general, the notion of a system is introduced to denote the existence of relationships between the data or variables being observed or more specifically, the existence of a transformation of some set of data into another set. The notion of a system in a certain sense can be considered to be a counterpart to the notion of a physical object and just as different physical theories deal with different kinds of physical objects, various aspects of information-processing and decision-making theories deal with various types of "systems," their behavior, transformation, control, etc. A theory of information processing and decision making can be considered therefore to be a theory of systems.

In this paper we shall be concerned with the mathematical theory of general systems. Essentially, a mathematical theory of general systems is a theory of mathematical models of real-life systems (in particular, described as information-processing or decision-making systems) such that the most basic properties of these systems are formalized by models which use minimal mathematical structures (compatible with the intuitive interpretation of the properties of concern). Such a theory is at present under development in the framework of set theory and the bordering branches of abstract mathematics and we shall highlight only some of the current developments. The material presented is based on earlier publications, and, in particular, on Mesarović (1964, 1967a, b, 1966, and 1968a)

II. FOUNDATIONS AND MOTIVATION

A fundamental concept in mathematical systems theory is the notion of a general system S, defined as a relation on abstract sets

$$S \subset V_1 \otimes \cdots \otimes V_n$$

Coordinate sets $V_i, \cdots, V_n$ of a system S are referred to as systems objects. In interpretation, each V_i represents the totality of all appearances of (or experiences with) an attribute of the real-life phenomena under consideration. Similarly, S represents the totality of all appearances of (experiments with) the real-life system.

In the further development in this paper, we shall assume that the family of objects $\bar{V}$ is partitioned in two classes $V_x = \{V_i \colon i \varepsilon I_x\}$ and $V_y = \{V_i \colon i \varepsilon I_y\}$ so that S can be represented as a binary relation

$$S \subset X \otimes Y$$

where X and Y are sets such that $X \leftrightarrow \otimes \{V_i \colon i \varepsilon I_x\}$, $Y \leftrightarrow \otimes \{V_i \colon i \varepsilon I_y\}$. X is called the input object while Y is called the output object. We shall be concerned with the binary relation representation of the system.

To illustrate the relationships between the notions of a system and the traditional ways of representing a model by means of a set of equations, consider two examples.

1. Let S be a system defined on two objects X and Y; furthermore, let two arbitrary finite sets A and B be given such that X is the family of all finite sequences of A while Y is the set of all finite sequences of B.

Let $S \subset A^T \otimes B^T$ be such that for any $(x,y) \varepsilon S$ both x and y are sequences of the same length. The system S is then a set of pairs of sequences on A and B, respectively, with the same length.

Consider now an automaton defined as a quintuple (A,B,Z,P,G) such that:

A is the set of input symbols
B is the set of output symbols
Z is the set of states
P is the next state function $P\colon A \otimes Z \rightarrow Z$
G is the output function

$$G\colon Z \otimes A \rightarrow B$$

System S, apparently, is a set of pairs of sequences of input and output symbols generated by an automaton by the application of P and G on a given input sequence.

2. Let A and B be reals and let X, Y be families of continuous differentiable functions defined on a time interval. A differential equation

$$\frac{dy(t)}{dt} = f(y(t), x(t))$$

can then be used to specify constructively a system $S = X \otimes Y$ such that for any given input function, $x(t)$, and the initial condition, the associated output represents the solution of the differential equations.

There are two ways in which the notion of a general system can be specialized: (1) by introducing additional structure—algebraic or topological—in the objects X and Y; (2) by representing the elements of X and Y as sets and introducing the structure in them. In the first case we shall talk about algebraic or topological systems while in the second case we shall talk about family systems. Of course, there is an intimate relation between two types of systems (e.g., certain types of family systems possess algebraic structures that are very convenient for the introduction of some auxiliary functions) and it is often a question of convenience whether algebraic or family structure is emphasized. We shall be concerned here with algebraic and family systems. The topological systems will not be considered to any extent.

Before proceeding with the formal development, let us briefly comment on the motivation for the initiation of this particular type of general systems theory and on the potentials for the application of the theory.

The mathematical theory of general systems is being developed to deal with, among others, the following main problems.

A. Structural Problems

One of the most crucial steps in the engineering process is selecting a structure for the system to be designed or, similarly, analyzing the consequences of the restructuring of a system. A detailed mathematical model, even when available, is not suitable for this purpose. Traditionally, engineers have used the block diagram basically for the purpose of grasping the overall composition of the system and for the subsequent structural considerations. The principal attractiveness of the block diagrams is their simplicity; their major drawback is the lack of precision. General systems theory can be an improved tool in basic structural considerations since it preserves the simplicity of the block diagrams while introducing the precision of mathematics. Actually, the role of general systems theory in the engineering methodology can, perhaps, best be represented by the diagram in Figure 1.

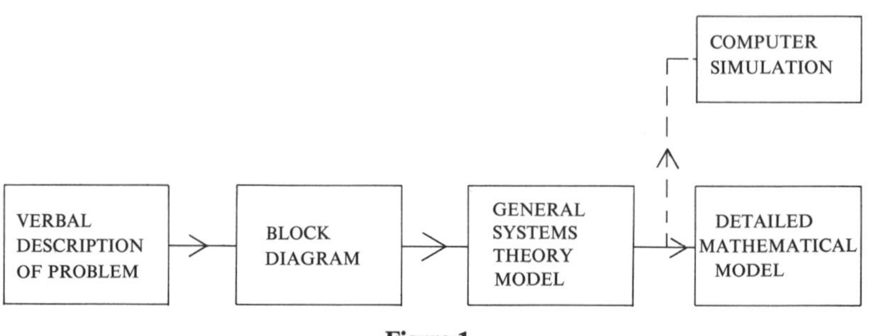

Figure 1

General systems models therefore lay between the block diagram representation and a detailed mathematical (or computer) model. In particular, in the study of large-scale complex systems, a general systems model represents a necessary step since the gulf between the block diagram and the detailed model can be too great. In addition, the availability of certain general systems techniques to treat the problem (at least in a preliminary way) on the general systems level can significantly enhance the usefulness of adding this step in the process.

B. Theory of Large-Scale Systems

Traditionally, problems of large-scale systems have been approached only after the system has been approximated by a sufficiently simplified model so that more conventional types of mathematical structures can be used in the modeling. Recently, two new approaches have been under active investigation: the *abstraction approach,* and the *hierarchical or multilevel approach.*

In the approach via abstraction one uses a mathematical model which is less structured and models only some of the dominant, "key," features of the problem. For example, suppose the system is described by a large number of partial differential equations. The study of the stability of such a system via the Liapunov method can be quite complex. However, if one recognizes the algebraic structure of the system's transformation, one might consider the stability problem algebraically (Windeknecht and Mesarović, 1967); thereby using a less detailed representation of the system. This approach is predicated on the development of a mathematical theory of general systems. The distinction between the approximation and abstraction approach should be noted. In the former, one uses the same mathematical structure and the simplification is achieved by the omission of some parts of the model that are considered as less important, e.g., a fifth-order differential equation is substituted by a second-order equation by considering only the two "dominant" state variables of the system. In the latter approach, however, one uses a different mathematical structure, one which is more abstract, and considers the system as a whole, but from a less detailed viewpoint. The simplification is not achieved by the omission of the variables but by the suppression of some of the details considered unessential.

In the hierarchical approach one decomposes the problem and solves the subproblems independently. Partial solutions are then coordinated by a hierarchy of decision processes aimed at coming as close to the overall solution as possible. The decomposition and the coordination can be done either in time or in space. In the design or computational problems, one solves the subproblems sequentially in time and achieves the coordination by an iteration process. In the problems of complex operational systems, the overall system is decomposed into subsystems whose operation is simultaneous in time and the coordination is achieved by "on-line" intervention during the actual operation of the subsystems.

It might be of interest to point out that, in spite of apparent dissimilarities between general systems theory and the theory of multilevel (hierarchical) systems, the motivation for their development is quite similar: to deal with complex, large-scale problems. Actually, one should look at these two theories as being intended to deal with the same type of problems but from different starting points. In Section VI we shall show how general systems theory can be used in structuring of a multilevel system.

C. Broadening Areas of Application of Mathematical Theories

Application of quantitative methods in many fields is often limited by the available mathematical models. Rather often the erroneous conclusions derived on the basis of some mathematical models are due primarily to the discrepancy between the structure of the actual phenomenon and the structure of the mathematical model used. It is the inadequacy of the particular model rather than the mathematical method per se that is at fault. General systems theory models can be very useful in such cases since they can correspond better to such real-life systems so that the discrepancy between the theoretical conclusions and the actual observations is smaller. The price paid for the improved accuracy is that such a general model has less information about the actual behavior of the real system.

D. Revelation of Interrelationships

A general systems theory can provide solid foundation and reveal the interrelationships between various specialized theories dealing with information processing and decision making. In actual real-life problem-solving situations, one often has to consider several alternative models for the description of the actual system, and the relationship between the results one can obtain by using different models can be best explored if all alternative models are presented in the same framework.

E. Precise Definition of Concepts and Interdisciplinary Communication

A general systems theory provides a framework for interdisciplinary communication since it is general enough so as not to introduce constraints of its own, yet, by its precision removes misunderstanding to a considerable degree. For example, different notions of adaptation used in the field of psychology, biology, engineering, etc., can be first formalized in general systems theory terms and then compared.

III. ALGEBRAIC STRUCTURES IN GENERAL SYSTEMS THEORY

The most immediate specialization of a general system is obtained by introducing algebraic structure in the objects X and Y. There are two principal objectives in introducing the algebraic structure. The first is to formalize the real-life systems concepts in the least constrained way and to investigate the most fundamental, structural, relationship between these concepts. In these studies, unusually weak algebraic structures (such as partial algebras or relational structures) and categorical algebra are used. Here we shall primarily deal with partial algebras. The second objective is to investigate properties of some special types of systems for which rather deep algebraic structures offer a good basis for modeling. For example, this is the case in the study of finite automata.

We shall be concerned primarily with the first objective since it is more germane to the general systems theory. However, in Section III, B, we shall also comment on the limitations which particular axioms of deeper algebraic structures introduce in the modeling of real-life systems.

A. Partial Algebras and General Systems Theory

To illustrate how the partial algebraic structures can be used in general systems theory, we shall consider the problems of state, state determinacy, and related concepts. This is a particularly natural domain for the use of partial algebraic structures. For example, the concept of state can be introduced solely by using the notion of relative product while the concept of state space can be introduced by using only two relations: relative product and an equivalence relation.

It should be noticed that many other problems in general systems theory can be treated satisfactorily by using (partial) algebraic methods. For example, in Windeknecht and Mesarović (1967) the problem of finite stability has been considered by using such structures.

1. Concept of State

Conceptual considerations indicate that the concept of state should be defined in terms of the following properties.

CONNECTIVITY. Assume that one considers only a subset $S'' \subset S$. The subset explicitly under consideration is usually referred to as "a part" or a "subsystem" or "incomplete observation" of S. (There are many different ways in which the notion of subsystem, or "part" can be defined more precisely. For our present purpose it is sufficient to consider a subsystem as a subset of S.) To account for the omitted part, one introduces the state and the subsystem is then connected via states with the rest of the system in such a fashion that the effect of the overall system on the subsystem under consideration is reflected correctly. All the information one should have about the rest of the system is contained in the state.

INVARIANCE. The state should be defined for the system as a whole rather than differently for each of the subsystems; the behavior of the entire system can be then studied in reference to a single "invariant" state object.

FUNCTIONALITY. For any given state, the input to a subsystem should uniquely define its output. In this way the "causality" of the system behavior is preserved.

The concept of state will be introduced in this paper so that it always has the property connectivity, preferably it has the property invariance, and hopefully it also has the property functionality. More specifically, connectivity will lead to the concept of a state object, the property invariance (in conjunction with connectivity) to the concept of a state space, while functionality (in conjunction with connectivity) will yield either a proper state object or the concept of a state-determined system.

a. GLOBAL STATE OBJECTS. The most immediate notions of state objects are introduced with reference to the connectivity and functionality properties and for the system S in its entirety, i.e., *without requiring any additional structure either for S or for the system objects $V_1,\cdots,V_n$.* We shall consider the binary form of the system, i.e., $S\subset X\otimes Y$.

i. Global state object. The notion of connectivity is formalized most conveniently by the concept of the relative product. Given two binary relations $R_1\subset A\otimes B$ and $R_2\subset C\otimes D$, if $B=C$, the composition $R=R_1\circ R_2$ is called the relative product of R_1 and R_2. A relative product is therefore defined only if the image of R_1 is equal to the domain of R_2. Conversely, given a relation $R\subset A\otimes B$, there exist, in general, many sets C such that there can be defined two relations $R_1\subset A\otimes C, R_2\subset C\otimes B$, and $R=R_1\circ R_2$. A relation R can be represented as a relative product in many different ways.

Consider the system $S\subset X\otimes Y$. Let Z be an arbitrary set such that there is given a pair of relations $S_Z\subset X\otimes Z$ and $S_Y\subset Z\otimes Y$ such that S is the relative product of S_Z and S_Y, $S=S_Z\circ S_Y$. The set Z represents the global state object for S while the elements of Z are global states.

By means of the global state object, the system S is decomposed into a series composition of two systems S_Z and S_Y. S_Z will be referred to as the input-state system while S_Y is the state-output system. The following proposition is obvious:

Proposition 1. Any system $S\subset X\otimes Y$ has a global state object, i.e., can be decomposed into S_Z and S_Y, $S=S_Z\circ S_Y$.

ii. Initial global state object. If we assume that the system under consideration represents a subsystem of a larger system, it is justified to introduce a state object so that the effect of the rest of the system is described. We have then

$$S_{Z_0} \subset Z_0 \otimes X \otimes Y$$

Preferably Z_0 should be such that some characteristic property of the overall system is preserved. In particular, if one assumes that the overall system was a function mapping inputs into outputs, one is justified in introducing Z_0 such that S_{Z_0} is a function

$$S_{Z_0}: (Z_0 \otimes X) \rightarrow Y$$

If S_{Z_0} is a function, Z_0 is referred to as the initial global state object.

The parentheses in the domain of S_{Z_0} are used as a notational convenience to indicate that S_{Z_0} is defined on a subset of $Z_0 \otimes X$. This convention will be used throughout this paper.

We have now the following:

Proposition 2. Every system $S \subset X \otimes Y$ has an initial state object.

Proof. Mesarović (1968a)

From Prop. 1 and Prop. 2 there follows immediately.

Corollary 1. Every system $S \subset X \otimes Y$ can be decomposed into two functional systems

$$S_{Z_0}: (Z_0 \otimes X) \rightarrow Z \text{ and } S_Y: Z \rightarrow Y$$

For each $z \varepsilon Z_0$ let $f_z: (X) \rightarrow Y$ such that $f_z(x) = S_{Z_0}(z,x)$. The system S can be then represented as a set of functions $S = \cup F$ where $F = \{f_z : z \varepsilon Z_0\}$. This will be referred to as functional representation of S.

b. STATE OBJECTS OF TRANSITIONAL SYSTEMS. A more frequently used notion of state refers to the succession of the systems appearances. To formalize the notion of succession we shall introduce algebraic structure in S.

i. State object. Let S be a partial algebra with the binary operator $C: (S \otimes S) \rightarrow S$. The operator C introduces minimal structure needed to formalize the notion of "succession" or "transition." The most natural interpretation of the operator C is for the case of time systems where C represents the concatenation of time functions. However, in general, C does not have to have all the properties needed by a concatenation operation; e.g., C does not have to have either left or right cancellation law. Indeed, the concepts of state object and state space will be introduced without qualifications regarding C.

Let $s, s'' \varepsilon S$ be such that $C(s,s'') \varepsilon S$. Specialization of C on s defines an elementary (right) translation $C_s:(S) \rightarrow S$ such that $C_s(s'') = C(s,s'')$ whenever defined. Similarly, if $s \varepsilon S$ and there exists $s' \varepsilon S$ such that $C(s',s) \varepsilon S$, then by specialization on S one obtains the (left) translation $C^s:(S) \rightarrow S$ such that $C^s(s') = C(s',s)$ whenever defined. These translations define two functions: the successor function $F:(S) \rightarrow \pi(S)$ such that

$$F(s)=C_s(S)=\{s'':C(s,s'')\varepsilon S\}$$

and the predecessor function $G:(S)\to\pi(S)$ such that

$$G(s)=C_s(S)=\{s':C(s',s)\varepsilon S\}$$

We can now introduce the concept of a state object and of a state space in such a general setting.

Denote by S_F the set of all elements of S which follow (in the sense of C) some elements in S, i.e.,

$$S_F=\{s:(\exists s'): C(s',s)\varepsilon S\}$$

Let S'' be an arbitrary subset of S, $S_F''=S''\cap S_F$, and S' the set of all elements in S which precede some elements in S_F'', $S'=G(S_F'')$. Define the relation

$$S^*\subset S'\otimes S_F''$$

such that for all $s'\varepsilon S'$ and $s''\varepsilon S_F''$

$$(s',s'')\varepsilon S^*\leftrightarrow C(s',s'')\varepsilon S$$

Let Z be an arbitrary set and $S_Z''\subset Z\otimes S''$ a relation such that there exists a subset $Z^*\subset Z$ and a pair of relations $S'^Z\subset S'\otimes Z^*$, $S_{ZF}''\subset Z^*\times S_F''$ such that S^* is the relative product of S'^Z and S_{ZF}''

$$S^*=S'^Z\text{o}S_{ZF}''$$

The set Z represents the state object for S'' and the elements of Z represent states. If S_Z'' is a function such that

$$S_Z'': (Z\otimes X'')\to Y''$$

Z will be referred to as a proper state object for S'' and the function S_Z'' as a state representation for S''. Then, the state object satisfies the connectivity, while the proper state object satisfies both the connectivity and the functionality.

ii. State space. To introduce the notion of a state space we have first to define an equivalence between states. Let $\bar{Z}$ be the union of all state objects defined for S, i.e.,

$$\bar{Z}=\cup\{Z'':[(\exists S'')\ S''\subset S]\wedge[Z''\text{ is a state object for }S'']\}$$

Let $E_z\subset\bar{Z}\times\bar{Z}$ be an equivalence relation. The quotient set

$$Z=\bar{Z}/E_Z$$

represents the state space for S.

Let S', $S''\varepsilon\pi(S)$ and Z'' be the state object that connects S' and S'', i.e., $S'^Z\subset S'\times Z$, $S_Z''\subset Z\otimes S''$ and $S'^Z\circ S_Z''\varepsilon\pi(S)$ (Z' is not necessarily a subset of the state space). There exists then a pair of mappings $A:Z''\to\pi(S'')$ and $B'':Z''\to\pi(S')$ such that for all $z\varepsilon Z''$

$$A(z)=\{s'': (s'\varepsilon S')\wedge(s''\varepsilon S'')\wedge C(s',s'')\varepsilon S\}$$

and

$$B(z)=\{s': (s'\varepsilon S')\wedge(s''\varepsilon S'')\wedge C(s',s'')\varepsilon S\}$$

A and B will be referred to as the state assignment mappings. To every state $z\varepsilon Z''$,$A(z)$ gives the elements of S'' that can occur if the system is in the state z, while B gives all the elements of S' that can bring the system in the state z.

As a rule, the state space, i.e., the equivalence E_z is defined in reference to an equivalence in S, $E_s \subset S \otimes S$. Denote by C_x a binary operator in X, $C_x:(X \otimes Y) \rightarrow X$ obtained from S by projection; similarly, C_y is the operator in Y, $C_y:(Y \otimes Y) \rightarrow Y$ obtained by projection. Let there be given a congruence* $E_x \subset X \otimes X$, i.e., an equivalence that admits C_x and, furthermore, also a congruence $E_y \subset Y \otimes Y$ that admits C_y. The state space Z will be then termed compatible with E_x, E_y if for all $z, z' \varepsilon Z$

$$(x,x') \varepsilon E_x \wedge (z,z') \varepsilon E_z \rightarrow (y,y') \varepsilon E_y$$

where $(z,x,y) \varepsilon S_Z$ and $(z',x',y') \varepsilon S_Z$.

The classical interpretation of equivalence for the class of time systems is in terms of a so-called "shifting operation" (Windeknecht, 1967). Informally, two inputs are considered as equivalent if one of them is the initial segment for the other except for the shift in time. For the outputs, however, there are also some other classical interpretations of the equivalence, e.g., two outputs can be considered as equivalent if they have the same last value. The equivalences presently considered, however, are quite abstract.

Two state spaces, Z and Z', will be termed compatible if they are both compatible with the same pair E_x, E_y. The state space satisfies the connectivity and the invariance but does not necessarily satisfy the functionality.

2. Algebraic Characterization of State Determinacy

Using the concepts of state object and state space introduced in Section III, B, it is possible to introduce many other state-related concepts and prove theorems about their relationships and properties in similarly weak algebraic structure. We shall consider here only the concept of a state-determined system. A state-determined property is related to the evolution of the system or the transition of the system from one state to another. The most natural formalization of the state determinacy is therefore in the framework of time systems. However, it is possible to introduce the state determinacy in the weak algebraic structure assumed so far for the system and we shall proceed to do that. To simplify the matter somewhat we shall first introduce the notion of a transitional system.

Let C be a binary operator in S so that S is partial algebra and $\mathcal{C}:\pi(S) \otimes \pi(S) \rightarrow \pi(S)$ is an operator such that for any $S', S'' \subset S$

$$\mathcal{C}(S',S'') = \{s: s' \varepsilon S' \wedge s'' \varepsilon S'' \wedge s \varepsilon S \wedge s = C(s',s'')\}$$

A system S is a transitional system if there is given a partition $\mathcal{S} = \{S_t^{t'}\}$ of S such that $\mathcal{S}$ is a partial algebra with the operation $\mathcal{C}$.

* The notion of a congruence is used here for the partial rather than complete algebra, namely,
$(x',x^{*\prime}) \varepsilon E_x \wedge (x'',x^{*\prime\prime}) \varepsilon E_x \rightarrow (C(x',x''), C(x^{*\prime},x^{*\prime\prime})) \varepsilon E_x$
whenever defined.

Consider now an equivalence class $S_t^{t'} \varepsilon \mathcal{S}$. There are two state objects associated with $S_t^{t'}$: Z_t, which connects $S_t^{t'}$ with the preceding elements and $Z_{t'}$ which connects $S_t^{t'}$ with the succeeding elements. With every $S_t^{t'}$ there is defined a relation

$$P_t^{t'} \subset Z_t \otimes X_t^{t'} \otimes Z_{t'}$$

such that $(z_t, x_t^{t'}, z_t') \varepsilon P_t^{t'}$ if and only if there exists $s_{t'} \varepsilon S$ such that $C(s_t, s_{t'}) \varepsilon S$ and $s_{t'} \varepsilon A(z_{t'})$.

$P_t^{t'}$ is termed a state transition relation.

A system is termed state determined if for any $S_t^{t'} \varepsilon \mathcal{S}$, $P_t^{t'}$ is a function, $P_t^{t'}: (Z_t \otimes X_t^{t'}) \rightarrow Z_{t'}$.

We can now give several propositions related to the state determinacy:

Proposition 1. Let S be a transitional system and let C_y have a left cancellation law. Then for any S^*, $S', S'' \varepsilon \mathcal{S}$ such that $S^* = \mathcal{C}(S', S'')$ if both $S_Z{}^*$ and $P_{t'}{}^{t''}$ are functions, $S_Z{}''$ is a function.

Proof. Assume that $S_Z{}''$ is not a function. There would exist then $z'' \varepsilon Z''$ such that $A(z'')$ is a relation, i.e., there exist $x'' \varepsilon X''$ and $y'', y''^* \varepsilon Y''$ such that $(x'', y'') \varepsilon A(z'')$ and (x'', y''^*). Let (z', x') be such that $P_{t'}{}^{t''}(z', x') = z''$. By the connectivity property of Z'' there exist $(x, y) \varepsilon S^*$ and $(x, y^*) \varepsilon S^*$ such that $x = C_x(x', x'')$ and $y = C_y(y', y''), y^* = C_z(y', y''^*)$. Since $S_Z{}^*$ is a function $y = y^*$. But then since C_y has left cancellation law, $y'' = y''^*$ and $S_Z{}''$ is a function.

Proposition 2. Let S be a transitional system such that for any $S_t^{t'} \varepsilon \mathcal{S}$, $S_{tz}{}^{t'}$ is a function, $S_{tz}{}^{t'}: (Z \otimes X_t^{t'}) \rightarrow Y_t^{t'}$. The system is then state determined.

Proof. Given $S^t{}_t{}' \varepsilon \mathcal{S}$, let $S_{t'}{}^{t''} \varepsilon \mathcal{S}$ be such that $C(S_t^{t'}, S_{t'}{}^{t''}) \varepsilon \mathcal{S}$. There exists then state-transition relation $P_t^{t'} \subset Z_t \otimes X_t^{t'} \otimes Z_{t'}$. Assume that $P_t^{t'}$ is a relation. There exist then $z_t \varepsilon Z_t$ and $z_{t'}, z_t^* \varepsilon Z_{t'}$ such that for given $x_t \varepsilon X_t^{t'}$, $(z_t, x_t, z_{t'}) \varepsilon P_t^{t'}$ and $(z_t, x_t, z_{t'}{}^*) \varepsilon P_t^{t'}$. If $A_x(z_{t'}) \cap A_x(z_{t'}{}^*) = \Phi$ for any such $z_{t'}$ and $z_{t'}{}^*$ we can put $z_{t'} = z_{t'}{}^*$ and $P_t^{t'}$ is a function. Assume therefore that there exist $x_{t'} \varepsilon A_x(z_{t'}) \cap A_x(z_{t'}{}^*)$. If for any such $x_{t'}$ there exists only one $y_{t'}$ such that $(x_{t'}, y_{t'}) \varepsilon S_{t'}$ we can again put $z_{t'} = z_{t'}{}^*$. Assume therefore that there exist $y_{t'}$ and $y_{t'}{}^*$ such that $(x_{t'}, y_{t'}) \varepsilon S_{t'}{}^{t''}$ and $(x_{t'}, y_{t'}{}^*) \varepsilon S_{t'}{}^{t''}$. If $y_{t'} \varepsilon A_y(z_{t'})$ and $y_{t'}{}^* \varepsilon A_y(z_{t'}{}^*)$, $S_{t'z}{}^{t''}$ is still a function. By the connectivity, however $(C_x(x_t, x_{t'}), C_y(y_t, y_{t'})) \varepsilon A(z_t)$ and $(C_x(x_t, x_{t'}), C_y(y_t, y_{t'}{}^*)) \varepsilon A(z_t)$. Since $S_{tz}{}^{t''}$ is a function, $C_y(y_t, y_{t'}) = C_y(y_t, y_{t'}{}^*)$. Due to the left cancellation law $y_{t'} = y_{t'}{}^*$. Therefore, we can set $z_{t'} = z_{t'}{}^*$ and $P_t^{t'}$ is a function.

An element $s \varepsilon S$ will be termed an initial element if there is no element $s^* \varepsilon S$ which precedes it, i.e., for which $C(s^*, s)$ is defined. Similarly, an equivalence class $S_t^{t'} \varepsilon \mathcal{S}$ will be called initial if there is no other class $S_{t'}{}^{t''} \varepsilon \mathcal{S}$ such that $C(S_{t'}{}^{t''}, S_t^{t'})$ is defined.

Given two classes $S_t^{t'}, S_t^{t''} \varepsilon \mathcal{S}$, $S_t^{t'}$ will be termed the initial segment of $S_t^{t''}$ if there exists $S_{t'}^{t''}$ such that $S_t^{t''} = \mathcal{C}(S_t^{t'}, S_{t'}^{t''})$.

We have now the following:

Proposition 3. Let S be a transitional system such that for any initial segment $S_{t_0}^{t} \varepsilon \mathcal{S}$, $S_{t_0 z}^{t}$ is a function $S_{t_0 z}^{t}$: $(Z \otimes X_{t_0}^{t}) \rightarrow Y_{t_0}^{t}$. Then if S is state determined, for any $S_t^{t'} \varepsilon S$, $S_{tz}^{t'}$ is a function $S_{tz}^{t'}$: $(Z \otimes X_t^{t'}) \rightarrow Y_t^{t'}$.

Proof. If $S_t^{t'}$ is an initial segment, $S_{tz}^{t'}$ is a function by assumptions. Assume that $S_t^{t'}$ is not initial. There exists then $S_{t_0}^{t}$ such that $S^* = C(S_{t_0}^{t}, S_t^{t'}) \varepsilon \mathcal{S}$. But then S^* is initial and S_z^* is a function. Since S is state determined, by Proposition 1 $S_{tz}^{t'}$ is a function $S_{tz}^{t'}: (Z \otimes X_t^{t'}) \rightarrow Y_t^{t'}$.

Proposition 4. Let S be transitional and state determined; then for any $S_t^{t'}$ and all $z_t, z_{t'}$, $A(z_t) \cap B(z_{t'})$ is a function.

Proof. Assume that $A(z_t) \cap B(z_{t'})$ is not a function. There would exist then $x_t^{t'}$ such that $P_t^{t'}(z_t, x_t^{t'}) = z_{t'}$ but there are two outputs $y_t^{t'}$ and $y_t^{t'*}$ such that $(x_t^{t'}, y_t^{t'}) \varepsilon B(z_{t'}), (x_t^{t'}, y_t^{t'*}) \varepsilon B(z_{t'})$. Let $S_{t'}^{t''} \varepsilon \mathcal{S}$ be such that $S_t^{t''} = \mathcal{C}(S_t^{t'}, S_{t'}^{t''}) \varepsilon \mathcal{S}$. Due to the connectivity property of $Z_{t'}$ there exists then $(x_{t'}^{t''}, y_{t'}^{t''}) \varepsilon S_{t'}^{t''}$ such that $(z_t, C_x(x_t^{t'}, x_{t'}^{t''}), C_y(y_t^{t'}, y_{t'}^{t''})) \varepsilon S_{tz}^{t''}$ and $(z_t, C_x(x_t^{t'}, x_{t'}^{t''}), C_y(y_t^{t'*}, y_{t'}^{t''})) \varepsilon S_{tz}^{t''}$. By Proposition 3, $S_{tz}^{t''}$ is a function, hence $C_y(y_t^{t'}, y_{t'}^{t''}) = C_y(y_t^{t'*}, y_{t'}^{t''})$. But C_y has a cancellation law and $y_t^{t'} = y_t^{t'*}$ and $A(z_t) \cap B(z_{t'})$ is a function.

Propositions 1 to 4 indicate two important characteristics of the state-determined systems: (1) Given a state and the future input, the future output is defined uniquely; the future behavior of the system is therefore uniquely defined whenever a state is given. (2) Let $S_t^{t'} \varepsilon \mathcal{S}$ and $Z_t, Z_{t'}$ be two associated state objects. For any given $z_t \varepsilon Z_t$ states $z_{t'} \varepsilon Z_{t'}$ represent equivalence classes in $X_t^{t'}$. Namely, to every $z_{t'} \varepsilon Z_{t'}$ corresponds an equivalence class $X(z_t, z_{t'}) \subset X_t^{t'}$ such that $x \varepsilon X(z_t, z_{t'})$ if and only if $P_t^{t'}(z_t, x) = z_{t'}$. If $E_{tx}^{t'}$ represents equivalence that generates that partition, the state object $Z_{t'}$ is the quotient set $X_t^{t'} / E_{tx}^{t'}$.

The generality of the concepts such as state objects, state space, state determinacy, etc., as well as the characteristic relationships established between these concepts should be noticed. No structure is needed to introduce the global and initial state object while only partial algebra is needed to introduce state object and state transition relations. For the state space and state determinacy in addition to the partial algebra the equivalence between states and a partition of S is all that is necessary. The conclusions that can be drawn about the existence of these characteristics reveal therefore the most essential conditions which the systems structure should possess. For example, the state determinacy of a system depends upon the equivalence that is used to compare different states. If no conditions are imposed on the state equivalence, for every system there exists a state space so that the system is state determined; furthermore, given the states for the initial equivalence class $S_{t_0}{}^{t} \varepsilon \mathcal{S}$, the state object for all other classes $S_t{}^{t'} \varepsilon \mathcal{S}$ can be obtained as the quotient set of the input objects.

Many other properties can be discussed on a similarly general level. The state concepts are selected to be discussed here primarily because the same concepts have been discussed in a number of different specific systems studies. For example, the recognition of the states as the equivalence classes on inputs is customary for systems defined on denumerable objects such as (finite) automata, systems described by difference equations, etc. On the other hand, state determinacy is an important concept for the systems specified by differential equations. The treatment of these concepts on the general level clearly indicates they do not depend upon special conditions such as cardinality of objects, differentiability or linearity conditions, etc., but rather on basic structural properties of systems.

Many other systems concepts can be introduced in weak algebraic structure. However, the establishment of the relationship between these concepts is rather complex since many special conditions have to be observed. It is therefore more efficient to introduce additional structure into the systems objects and to investigate the systems property in a more regular setting.

B. Algebraic Theory of Systems

Algebraic structures are used primarily in the theory of finite state systems, so-called automata. We shall indicate what assumptions should be added to the notion of general system in order to develop a similar algebraic formalism. We shall place the emphasis on the structural properties which the system object should have, regardless of their cardinality. Furthermore, we shall briefly indicate what are the benefits and limitations which any of these assumptions introduce.

1. Closure

The most essential assumption for the algebraic theory of systems is that the objects X and Y are full algebras rather than partial algebras as we have assumed for general systems in Section III, A. The binary operation C_x is then $C_x : X \otimes X \rightarrow X$ and similarly $C_y : Y \otimes Y \rightarrow Y$. While the closure property of (full) algebras is a basic property needed for many rather deep results of the algebraic theory of systems, many systems of considerable practical interest fail to have this property. For example, let both X and Y be sets obtained by restrictions from a space of continuous, differentiable functions (such that the system can be specified by a set of ordinary differential equations) and, furthermore, the algebraic operation C is such that for any two elements of X(or Y) the value of C is obtained by shifting and concatenation of the argument functions. The objects X and Y cannot be then closed since the application of C, in general, yields functions with discontinuities.

It should be noticed that this is not the case if the system is defined on discrete time, e.g., when a set of discrete equations is used for systems specification. In this rather fundamental respect, these systems are closer to the automata type systems than to the systems specified by differential equations.

2. Associativity

In addition to closure, the algebraic operation is required to be associative, i.e., for any $x,x',x'' \varepsilon X$, $C_x(x,C_x(x',x''))=C_x(C_x(x,x'),x'')$. The algebra is then a semigroup. If the objects are algebras rather than partial algebras, associativity appears as a very natural property. In particular, if the system is state determined, the associativity of the input object is needed for the composition of the state-transition functions, $P_{t'}{}^{t''}$, such that, for any $z \varepsilon Z_t$, $x_t{}^{t'} \varepsilon X_t{}^{t'}$ and $x_{t'}{}^{t''} \varepsilon X_{t'}{}^{t''}$ the following identity holds.

(*) $P_t{}^{t''}(z_t,C_x(x_t{}^{t'},x_{t'}{}^{t''}))=P_{t'}{}^{t''}(P_t{}^{t'}(z_t,x_t{}^{t'}),x_{t'}{}^{t''})$

Condition (*) is referred to as semigroup property of the state transition functions.

However, for a large class of time-varying systems, the family of state transition functions does not have semigroup property.

3. Generating Sets and Functions

The most natural way to define an algebra constructively is by means of the generating set. Indeed, any algebra X has a generating set A such that any element of X is obtained by the repeated applications of the operator starting from the elements of A. Furthermore, for any algebra X there exists a word-algebra X_w, with the generating set A_w such that X is a homomorphic image of X_w. A word algebra has a binary operator such that any $x_w \varepsilon X_w$ is a sequence of elements from the generating set A_w. To any algebra therefore there can be given a word algebra X_w, with A_w such that to every $x \varepsilon X$ there could be assigned a sequence of elements from A_w. Elements of X can be then identified by the sequences from A_w.

A standard assumption for the algebraic theory of systems (automata) is that the objects are word algebras, i.e., any $x \varepsilon X$ is a sequence of generators. If A is the generating set and N is the set of natural numbers $x \varepsilon X$ is then $x:N' \rightarrow A$ where $N' = \{0,1,\cdots,n'\}$. If $F=A^N$ denotes the set of all functions on N into A, X is then obtained by taking restrictions from A^N. It should be emphasized that the generator set is actually the image for the function set F that is used for the construction of X.

Consider now the set of functions on reals R into a set B, $F=B^R$. An algebraic object X can again be constructed from F by taking restrictions of the elements $f \varepsilon F$ and by concatenating them. Such an object again will have the set of generators A_w but it is now different than the codomain of the functions from F, i.e., $A_w \neq B$. Indeed, the generator set A_w is a family of restrictions of $f \varepsilon F$ such that, again, any $x \varepsilon X$ is a sequence of generators, i.e., x: $N' \rightarrow A_w$. However, none of the elements of X can be obtained algebraically from B, i.e., by repeated application of C.

The role of the generating set in general systems theory is greatly changed because of this distinction, with a considerable loss in simplicity of description and the usefulness of the notion in itself. Similarity between systems specified by difference equations and automata should be noticed. The identity of the generating set and the range of the inputs is preserved and the algebraic approach used in automata theory can be easily extended to the difference equations systems. The difficulty, of course, begins when the index set is real line or has even higher cardinality. This aspect of the algebraic approach is not therefore applicable to differential equation systems.

Let us mention in passing that the algebraiclike treatment of systems of arbitrary cardinality would require a generalization of the notion of algebra. Traditionally, an algebra is defined in terms of an operation of finite order. Recently, Rasiowa and Sikorski (1966) have considered the notion of an algebra with the operation whose order is denumerable. To develop an algebraic approach to general systems, the notion of an algebraic operation must be generalized still further and the function must be allowed the arguments which are sets of infinite, nondenumerable, cardinality. The generalization here would not be unlike the generalization of the notion of a Cartesian product from the finite to arbitrary cardinality.

4. Cardinality

Finally, there are problems which are meaningful only in reference to the finitness of the objects of an automaton, for example, the problems of state assignment and minimal state space. The family of all state spaces for a finite automaton are ordered since they are finite sets of different cardinality and the procedure to find the state object of minimal cardinality is very important. On the other hand, for the systems in which any state space is infinite, the family of state spaces may not have a minimal element at all.

It should be noticed that there also exists a minimization problem associated with the construction of the state space for the systems which have the space of infinite cardinality. That is the problem of finding the minimal number of state variables for a system specified by differential equations. But this is a completely different problem than the minimal state space problem in automata theory. It refers to finding how to represent a state as a sequence of a minimal number of given elements. For example, the state can be represented by a pair of real numbers rather than a quintuple of reals. The state space Z is essentially represented as a Cartesian product $Z = Z_1 \otimes \cdots \otimes Z_n$ and the problem is to find the representation of Z in terms of minimal number of components. The cardinality of the state space Z does not have to be changed at all. In the automata theory, however, the minimization of the state space refers to the minimization of the cardinality of Z itself.

IV. GENERAL TIME SYSTEMS

A. State Objects and State Space

The notion of a time system is obtained by specializing the systems objects. In the algebraic approach additional structure is introduced in S or in X and Y. Each element $x \varepsilon X$ and $y \varepsilon Y$ is still considered an indivisible entity but there exists a transformation between these elements. An alternative way to introduce structure in the objects X and Y is to define its members, $x \varepsilon X$ and $y \varepsilon Y$ as sets.

A system object, V, is called a family object if each element $v \varepsilon V$ is a set. Furthermore, V is termed a function-generated object if each v is a function $v: T_v \rightarrow A_v$. T_v will be referred to as the index set for v while $A = \cup \{A_v : v \varepsilon V\}$ will be referred to as the alphabet* for V. The alphabet for an input object will also be referred to as the input space while the alphabet for the output will be called the output space. Often, a function-generated object is obtained by some suitably chosen restrictions from a family of functions with the same domain and codomain.

Further specialization of objects is obtained by structuring the index set. A linearly ordered index set is called a time set. A function-generated object defined on time sets is called a time object.

Consider a system $S \subset X \otimes Y$. S is termed a family system if both X and Y are family objects. S is a time system if both X and Y are time objects.

* Notice that the notion of an alphabet is nonstandard since A_j can be of arbittrary cardinality.

In the subsequent development we shall be concerned solely with the time systems of the following form

$$S \subset A^T \otimes B^T$$

Such a system will be referred to as a complete time system. An incomplete time system can be obtained from S by restrictions on the appropriate index sets.

Finally, we shall need the notion of a system generated from a complete system by taking the restrictions on appropriate subsets. To that end, the following subsets of the time set will be used: For any $t \varepsilon T$,

$$T^t = \{t^*: t < t^*\} \text{ and } T_t = \{t^*: t^* \leq t\}$$

For any pair $t, t' \varepsilon T$, $t < t'$, $T_t^{t'} = \{t^*: t < t^* \leq t'\}$. Restrictions of the system on these subsets will be then defined: Namely, if $X^t = X \mid T^t$ and $Y^t = Y \mid T^t$ then $S^t \subset X^t \otimes Y^t$, such that $S^t = S \mid T^t$, i.e.,

$$(x^t, y^t) \varepsilon S^t \leftrightarrow \exists (x,y): (x,y) \varepsilon S \wedge x^t = x \mid T^t \wedge y^t = y \mid T^t$$

$S_t = S \mid T_t$ and $S_t^{t'} = S \mid T_t^{t'}$ are defined similarly.

We can now introduce the global system $\bar{S}$ generated by S

$$\bar{S} = \cup \{S_t^{t'}: t, t' \varepsilon T \wedge t < t'\}$$

For any $t \varepsilon T$ the successor function F and the predecessor function G take on the form

$$F_t: S_t \rightarrow \pi(S_t)$$

such that

$$F_t(s_t) = \{s^t: (\exists s)(s_t = s \mid T_t \wedge s^t = s \mid T^t)\}$$

and similarly

$$G: S^t \rightarrow \pi(S_t)$$

such that

$$G_t(s^t) = s_t: (\exists s)(s_t = s \mid T_t \wedge s^t = s \mid T^t\}$$

B. State and State Determinacy for Time Systems

General time systems offer a most natural setting for various concepts of state and state determinacy. We shall develop these concepts as special cases of the concepts introduced in Section III and will then proceed to investigate the conditions under which decomposition of systems by means of the global state object, i.e., $S = S_Z \circ S_Y$, take a specially convenient form.

We shall consider a global system generated by a complete time system $S \subset A^T \otimes B^T$.

Let $t \varepsilon T$. The state object at t is a set Z_t such that $S^* \subset S_t \otimes S^t$ is a relative product of $S_t^z \subset S_t \otimes Z_t$ and $S_z^t \subset Z_t \otimes S_t$, i.e.,

$$S^* = S_t^z \circ S_z^t$$

If S_z^t is a function

$$S_z^t: (Z_t \otimes X^t) \rightarrow Y^t$$

Z_t is termed the proper state object at t.

S_z^t will be referred to as the state representation for S^t.

For each state object Z_t there is given a pair of state assignment functions

$$A_t : Z_t \rightarrow \pi(S^t) \text{ and } B_t : Z_t \rightarrow \pi(S_t)$$

such that for all $z_t \varepsilon Z_t$

$$A_t(z_t) = \{s^t : (z_t, s^t) \varepsilon S_Z^t\} \text{ and } B_t(z_t) = \{s_t : (s_t, z_t) \varepsilon S_t\}$$

The state space for time systems will be also introduced as a direct specialization of the abstract notion of the state space from Section III. Let $\bar{X}$ and $\bar{Y}$ be the input and output objects of the global system $\bar{S}$ generated by $S \subset A^T \otimes B^T$, i.e.,

$$\bar{X} = \{X_t^{t'} : t, t' \varepsilon T \wedge t < t'\}$$
$$\bar{Y} = \{Y_t^{t'} : t, t' \varepsilon T \wedge t < t'\}$$

Define the equivalence relations $E_x \subset \bar{X} \otimes \bar{X}$ and $E_y \subset \bar{Y} \otimes \bar{Y}$. The input equivalence E_x is usually such that each equivalence class x^{E_x} is separated in $\bar{X}$, i.e.,

$$x, x' \varepsilon X_t^{t'} \wedge (x, x') \varepsilon E_x \rightarrow x = x'$$

Consider now the family

$$\bar{Z} = \cup \{Z_t : t \varepsilon T\}$$

Let the equivalence relation $E_z \subset \bar{Z} \otimes \bar{Z}$ be defined such that E_z is compatible with E_x and E_y, i.e., the states $z, z' \varepsilon \bar{Z}$ are equivalent, $(z, z') \varepsilon E_z$, if for all $(x, y) \varepsilon S_z^t(z)$ and $(x', y') \varepsilon S_z^{t'}(z')$

$$(x, x') \varepsilon E_x \rightarrow (y, y') \varepsilon E_y$$

The state space for a time system is then simply the quotient set

$$Z = \bar{Z} / E_z$$

When the state space for a time system is given, the state object for every $t \varepsilon T$ is a subset of Z. The state transition relation is then

$$P_t^{t'} \subseteq Z \otimes X_t^{t'} \otimes Z$$

It is easy to show that the state determinacy for the time systems is a specialization of the concept from Section III. Indeed, the global time system $\bar{S} = \{S_t^{t'} : t, t' \varepsilon T\}$ is a transitional system since any two members $S', S'' \varepsilon \bar{S}$ are nonintersecting, $S' \cap S'' = \Phi$. Furthermore, the concatenation of the time functions is an algebraic operation in both X and Y.

A number of theorems concerned with the existence of the state object, state space, state determinacy, etc., can be now derived either as an immediate consequence of the algebraic results from Section III or using the more specialized structure of the time systems. We shall only state some of them here. Proofs and more detailed discussion can be found in Mesarović (1968a).

Proposition 4. Let S be a time system. For every $t \varepsilon T$ there exists a proper state object at t, i.e., $S_z^t : (Z \otimes X^t) \rightarrow Y^t$.

Proposition 5. For any t, $t' \varepsilon T$, $t' > t$, there exists a pair of state objects $Z_t, Z_{t'}$ such that $P_t^{t'}$ is a function.

Proposition 6. Let $Z_t, Z_{t'}$ be the state objects such that Z_t is proper and $P_t^{t'}$ is a function. Then $Z_{t'}$ is also a proper state object.

Corollary 1. For any $t, t' \varepsilon T, t' > t$ there exists a pair of proper state objects $Z_t, Z_{t'}$ such that $P_t^{t'}$ is a function.

Proposition 7. For every time system S there exists a state space such that S is state determined.

Proposition 8. Let S be a time system with a state space Z. If $S_z{}^{t_0}$ is a function and S is state determined, then for all $t \varepsilon T$, $S_z{}^t$ is a function, $S_z{}^t: (Z_t \otimes X^t) \rightarrow Y^t$. Conversely, if for all $t \varepsilon T$, $S_z{}^t$ is a function, there exists a compatible state space Z such that S is state determined.

Proposition 9. If S is state determined, $A_t^{t'}(z) \cap B_t^{t'}(z')$ is a function for all $z \varepsilon Z_t$ and $z' \varepsilon Z_{t'}$ and for any t, t'.

Proposition 10. If S is state determined, $S_{tz}{}^{t'}$ is a function for any $t, t', t' > t$. Conversely, if $S_{tz}{}^{t'}$ is a function for all t, t' and the time set T has a maximal element, the system is state determined.

To investigate the fundamental decomposition $S = S_Z \circ S_Y$ more deeply we have to introduce new auxiliary functions.

For each $t, t' \varepsilon T$ the restriction $S_{tZ}{}^{t'}$ defines a relation

$$M_t^{t'} \subset Z_t \otimes X_t^{t'} \otimes Y(t')$$

such that

$$(z, x_t^{t'}, y(t')) \varepsilon M_t^{t'} \leftrightarrow (\exists y_t^{t'})((z, x_t^{t'}, y_t^{t'}) \varepsilon S_{tZ}{}^{t'} \wedge y(t') = y_t^{t'} \mid \{t'\})$$

$M_t^{t'}$ will be called a state-output transition relation.

For any $t' \varepsilon T$, the state relation $P_t^{t'}$ and the state-output transition $M_t^{t'}$

$$P_t^{t'} \subset Z_t \otimes X_t^{t'} \otimes Z_{t'}$$
$$M_t^{t'} \subset Z_t \otimes X_t^{t'} \otimes Y(t')$$

define a relation

$$G \subset Z_{t'} \otimes Y(t')$$

such that

$$(z', y\ (t')) \varepsilon Z_{t'} \otimes Y(t') \leftrightarrow (\exists x_t^{t'})(\exists Z)(z, x_t^{t'}, z') \varepsilon P_t^{t'} \wedge (z, x_t^{t'}, y\ (t')) \varepsilon M_t^{t'}$$

G is called the output generating relation at t. Characterization of G as a function is given by the theorem:

Proposition 11. Let S be a time system with a state space Z such that S is state determined and for each $t \varepsilon T$, $M_{t_0}{}^t$ is a function. There exists then a state space compatible with Z such that the diagram is commutative.

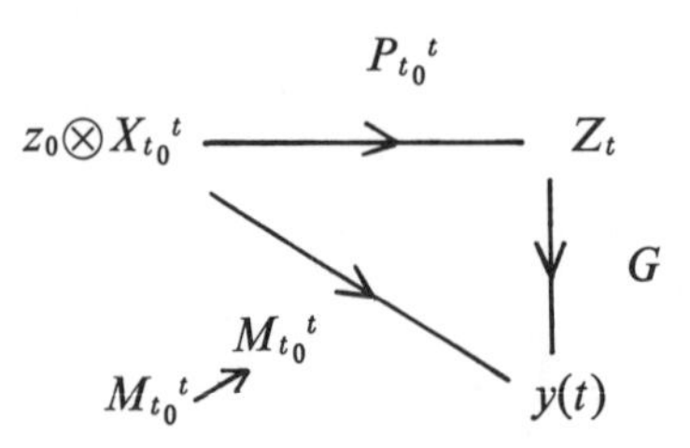

Proof. Let R be an equivalence in $Z_o \otimes X_{t_0}{}^t$ such that
$((z_0, x_{t_0}{}^t), (z_0{}^*, x_{t_0}{}^{t*})) \varepsilon R \leftrightarrow P_{t_0}{}^t(z_0, x_{t_0}{}^t) = P_{t_0}{}^t(z_0{}^*, x_{t_0}{}^{t*}) \wedge M_{t_0}{}^t(z_0, x_{t_0}{}^t) = M_{t_0}{}^t(z_0{}^*, x_{t_0}{}^{t*})$

We shall decompose R in two different ways in reference to the two conditions on the right side of the above equation.

First, consider the kernel of $P_{t_0}{}^t$, ker $P_{t_0}{}^t$, which is equivalent in $Z_0 \otimes X_{t_0}{}^t$. Apparently, R is a subset of ker $P_{t_0}{}^t$ since

$$(p,p') \varepsilon R \rightarrow (p,p') \varepsilon P_{t_0}{}^t$$

By the factor theorem for mappings there exists a unique function

$$F\colon Z_0 \otimes X_{t_0}{}^t / R \rightarrow Z_t$$

such that the diagram

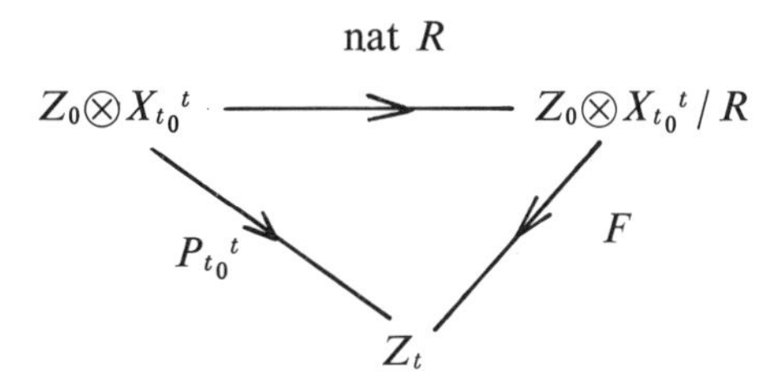

is commutative.

Second, by a similar argument $R \subset$ ker $M_{t_0}{}^t$ and by the factor theorem then there exists a unique mapping G

$$G\colon Z_0 \otimes X_{t_0}{}^t / R \rightarrow Y(t)$$

such that the diagram

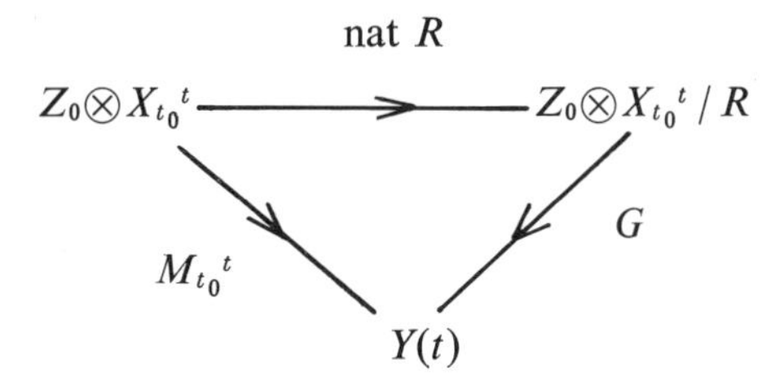

is commutative.

Let $Z_t{}^* = Z_0 \otimes X_{t_0}{}^t / R$. By Lemma 3 there exists a pair of state assignment functions so that $Z_t{}^*$ is a state space compatible with Z_t, since F is an onto function. Then nat $R = P_{t_0}{}^t$ and the theorem is proven.: :

Proposition 11 states an important property of the decomposition of the state-determined system. Namely, if the system is state determined, there exists a decomposition of S into S_t and S_y such that for any $t \varepsilon T$ the value of the output $y(t)$ is determined solely by the state $z(t)$ while the same state $z(t)$ is determined solely by the initial state $z(t_0)$ and the input $x_{t_0}{}^t$.

The usefulness of decomposition of S into S_Z and S_Y is primarily due to this property. The evolution of a state-determined system in time can be now studied solely in the state space Z since for any $z(t)$ the value of the output $y(t)$ is completely defined.

Notice the relation between the global state Z and the state space Z. Namely, if T is the time set $Z = Z^T$, the set of states (in the state space) for all $t \varepsilon T$ that corresponds to a given $(x,y) \varepsilon S$, i.e., $\{z(t) : t \varepsilon T\}$ is often referred to as a trajectory in the state space Z.

V. GENERAL THEORY OF GOAL-SEEKING SYSTEMS

There are two basic ways to provide a constructive specification of systems: the terminal approach and the goal-seeking approach.

1. The constructive specification of terminal systems is arrived at by providing additional structure on the object sets so that a simpler system can be defined (hopefully even of finite cardinality) that can be used to specify the original system, e.g., via a process of recursion or induction. Such simpler systems used for constructive specifications are called auxiliary functions (Mesarović 1966, 1968a). Often, they require introduction of some new (auxiliary) objects in the description of systems, most notably the state object or the state space.

2. Constructive specification in (and indeed the definition of the concept of) the goal-seeking approach is achieved by introduction of the notion of a goal for a system and then by describing the behavior of the system in reference to that goal. It should be emphasized that the goal-seeking description of a system is needed, not for philosophical or conceptual reasons, but rather for the purely technical reasons of arriving at a constructive specification. It might even be considered as an alternative way for implicit definition of a function (or relation); namely, for a given class of systems one might not have available a constructive specification via the terminal approach but only in terms of goal seeking. That does not mean, of course, that the basic concepts involved in the terminal description of such a system (e.g., the state object) cannot be defined; rather, it might be because the associated auxiliary functions are not available in an analytic or algorithmic form so that one has to resort to the goal-seeking approach (Mesarović 1968b).

The objective of this section is to formalize in general systems terms the notion of a goal-seeking system; this will be done in order to illustrate both the process of formalization and the kind of notions usually developed in general systems theory. No further formal development (propositions) will be presented here but the paper by R. Banerji illustrates how the approach to the goal seeking presented in this section can be used as a basis for the development of a theory of nonnumerical problem-solving situations.

Given a system

$$S \subset X \otimes Y$$

to arrive at the notion of a goal-seeking representation for S we need two preliminary concepts; namely, the concept of a goal and the concept of a decision maker. (For simplicity we shall view S as a function which implies that the members of X are input-state pairs.)

A. Goal

Let $X = M \otimes U$. A goal for S is defined then by a triplet of relations $\alpha = (G,T,R)$ defined in reference to a set V such that

$$G : S \rightarrow V$$
$$T : U \rightarrow V$$
$$R \subset V \otimes V$$

The set V represents the value or performance measure set. Under interpretation, G represents the performance (or goal) function that assigns a value $G(s) \varepsilon V$ to every appearance of the system, i.e., $s \varepsilon S$. T represents the tolerance (reference) function. For every $u \varepsilon U$, T gives the value $T(u) \varepsilon V$ that should be used to evaluate the performance of a given $y = S(m,u)$. Finally, R represents the satisfaction relation. For any $(m,u) \varepsilon M \otimes U$, the satisfaction with the behavior of the system will be evaluated with reference to $(G(m,u),S(m,u)),T(u)$ and R.

Given a goal $\alpha = (G,T,R)$ for a system S we have then two notions relating the inputs with the goal.

The input $x \varepsilon X$ achieves the goal α if

$$(G(x,S(x)),T(u)) \varepsilon R$$

where $x = (m,u)$.

The input $m \varepsilon M$ satisfies the goal α relative to $U' \sqsubset U$ if for all $u \varepsilon U'$ the input $x = (m,u)$ achieves the goal α, i.e., for all $u \varepsilon U'$

$$(G(m,u,S(m,u)),T(u)) \varepsilon R$$

The triplet $\beta(S,U'\alpha)$ will be referred to as a decision problem. An input $m \varepsilon M$ satisfies the decision problem (S,U',α) if it satisfies the goal α relative to U'.

B. Decision Maker (Decision System)

Given a system

$$S : M \otimes U \rightarrow Y$$

informally, the system will be referred to as the decision maker if a decision problem β is given such that for every $(m,u) \varepsilon M \otimes U$, the output $y = S(m,u)$ satisfies β (in a given sense).

More precisely, S will be termed a decision-maker if the following is given:

A pair of mappings

$$P: Y \otimes U \rightarrow M$$
$$W: U \rightarrow M$$

such that

$$m = W(u) \leftrightarrow (S(m,u) = y) \wedge (P(y,u) = \mathrm{m})$$

i.e., W is a composition of S and P as specified by Windeknecht and Mesarović (1967).

A goal α for P such that for all $u \varepsilon U$, $S(w, W(u))$ satisfies the decision problem $\beta = (P, U', \alpha)$ where $U' \subset U$ is defined by a predefined set-valued mapping F: $U \rightarrow \pi(U)$.

Under interpretation, U represents the uncertainty set and the mapping F selects (e.g., by prediction) the subset U' for which the output of S should achieve the given goal.

C. Goal-Seeking System

Finally, we are in a position to introduce the notion of a goal-seeking system.

Given a system

$$S: X \rightarrow Y,$$

there are two ways in which S can be defined as a goal-seeking system.

1. Let α be a goal for S. The system is considered as an (open-loop) goal-seeking system if every $x \varepsilon X$ satisfies the goal α.

2. S is considered as a (feedback) goal-seeking system if a set M is given together with a pair of mappings (D,P)

$$\mathrm{P}: M \otimes X \rightarrow Y$$
$$D: X \otimes Y \rightarrow M$$

such that

$$y = S(x) \leftrightarrow (P(m,x) = y) \wedge (D(x,y) = m)$$

D is a decision maker relative to a goal α for a mapping P_M defined on $M \otimes U$ into Y, i.e.,

$$P_M: M \otimes U \rightarrow Y$$

Apparently, according to the second notion, S is a goal-seeking system if there is given a pair of mappings (P,D) such that S is a (feedback) composition of P and D and, furthermore D is a decision maker relative to a goal α defined for P_M.

Starting from the notion of a goal-seeking system, some other notions can be defined, such as learning (adaptation), self-organization, etc. For example, learning can be defined as a process aimed at the reduction of the uncertainty set U (Mesarović, 1963, 1964*a*) while self-organization can be defined as a process of changing the structure of the goal-seeking process, i.e., the functions defining a goal-seeking system (such as performance function, process model P_M, tolerance function, satisfaction relation, etc.).

VI. GENERAL THEORY OF MULTILEVEL SYSTEMS

Last, but not least, let us consider how general systems theory can be used to solve structural problems in multilevel (hierarchical) systems.

A multilevel system is a system S such that: (1) S is specified as a family of interconnected subsystems. (2) Subfamilies of the subsystems are represented as decision makers (control, or goal-seeking subsystems). The subsystems which are not decision making represent the (basic) process. (3) The decision-making subsystems are arranged in a hierarchy such that some of the units do not have direct access to the basic process but rather receive the information and can influence only other decision units.

Multilevel systems are of considerable interest in many fields and are found in the form of social and economic systems (organizations), industrial automation, biological systems, etc. In particular, multilevel systems are of special importance in the large-scale systems area.

Before proceeding with the discussion of how general systems theory can help in the study of multilevel systems, let us point out that one of the principal reasons for introducing the extremely abstract notion of a general system (relation on abstract sets) is to be able to study systems that consist of large numbers of interconnected subsystems. If we are going to concentrate on the problems of interconnections it is essential that the building blocks (the subsystems) are described in a comparatively simple way, i.e., only those aspects of subsystems that are important for the interconnections and the behavior of the overall system are presented. For example, in some of the problems considered subsequently in this chapter, the monotonicity of certain functions, or the simultaneous attainment of optimality for all functions of a given family, are the only important properties. All other aspects of these functions (such as whether they are numerical or symbolic, continuous or discrete, etc.) are irrelevant and only make the essential study of the interdependence of the units across the level more difficult.

We shall consider more specifically the problem of structuring a two-level system.

The block diagram of the system under consideration is given in Figure 2.

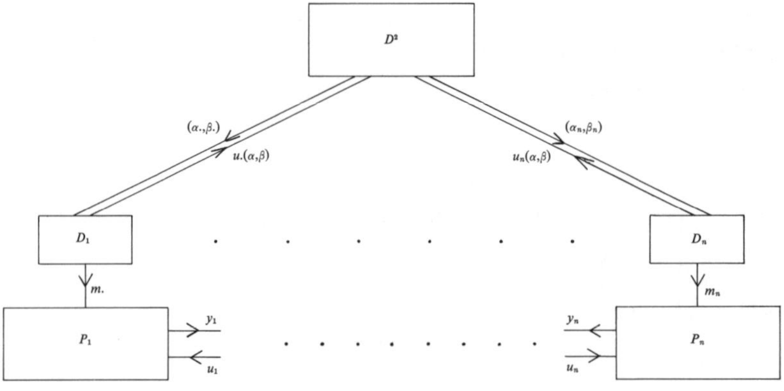

Figure 2

The system consists of a single decision unit D^2 on the second level and n units $D_1, \cdots, D_n$ on the first level. Assume that the decision problem Δ_i, of an arbitrary first level unit, D_i, is specified by the process

$$P_i: M_i \otimes A_i \rightarrow Y_i$$

and the performance function

$$G_i: M_i \otimes Y_i \otimes B_i \rightarrow V_i$$

where M_i is a (local) decision (control) object, Y_i the (local) output object, A_i and B_i are coordination objects while V_i is the evaluation object. The coordination objects A_i and B_i are specified by the second-level unit. The decision problem Δ_i for D_i is then to find the input $(\hat{m}_i, \hat{\alpha}_i) \varepsilon M_i \otimes A_i$ such that for all $(m_i, \alpha_i) \varepsilon M_i \otimes A_i$

$$G_i(\hat{m}_i, \hat{\alpha}_i) \leq G_i(m_i, \alpha_i)$$

It will also be assumed that there is given a decision problem Δ specified by the process

$$P: M \rightarrow Y$$

and the performance function

$$G: M \otimes Y \rightarrow V$$

such that

$$M = M_1 \otimes \cdots \otimes M_n \text{ and } Y = Y_1 \otimes \cdots \otimes Y_n$$

and the problem Δ consists in finding $\hat{m} \varepsilon M$ such that for all $m \varepsilon M$

$$G(\hat{m}) \leq G(m)$$

Δ will be referred to as the overall problem.

Decision problems Δ and $\{\Delta_i : \varepsilon N\}$ are related via the processes P and $\{P_i : i \varepsilon N\}$. Namely, it will be assumed that the projection of P_i on $M_i \otimes Y_i$ is a subset of the projection of P on the same objects $M_i \otimes Y_i$. In other words, the overall process P: $M \rightarrow Y$ is obtained by the interconnection of the processes $P_i^*: M_i \otimes U_i \rightarrow Y_i$ where the interconnection object U_i is a family of outputs from other subsystems. This will mean that the local process P_i is a correct model of the subsystem of P which is under local jurisdiction except for the interactions (which are not accounted for in P_i) and the coordination (which is absent in P, i.e., in its appropriate subsystem). No relation between the local and the overall performance functions will be assumed at this point.

The second-level unit has now a problem of coordination, i.e., it has to influence the first-level units $\{D_i : i \varepsilon N\}$ so that the solutions of the local problems $\{\Delta_i : i \varepsilon N\}$ will imply the solution of the overall problem Δ. A straightforward way to accomplish this would be to assign the overall problem Δ to the coordinator D^2. However, such an approach would almost eliminate the rationale for having a multilevel structure. To make the operation of the two-level system efficient it is necessary therefore to define a separate problem Δ^2 for the coordinator (which problem should be considerably simpler than Δ). The solution of Δ^2 will yield coordination such that the overall problem can be solved on the first level by the solution of the local problems. Specifically, in the context of the optimization problem considered in this paper, the coordination should be such that the solutions of the local problems $\{\Delta_i : i \varepsilon N\}$ yield such an input $\hat{m}$ to the process P that for all $m \varepsilon M$

$$G(\hat{m}) \leq G(m)$$

To synthesize the two-level system under consideration one has to be concerned with two aspects: (1) Structural: how to specify the problem Δ^2. (2) Procedural: after Δ^2 is specified, what are the effective procedures for solving Δ^2? We shall be concerned here only with the structural aspect.

A. Coordination Principles

In order to be able to proceed with the mathematical investigations of the second-level decision problem, some assumptions regarding the structure of Δ^2 will have to be made. From the structure of the first-level problems it is apparent that the decision objects for Δ^2 will have to be $A_i - s$ and $B_i - s$. However, nothing is implied about the problem Δ^2 itself. In such situations one has to start by postulating, on the intuitive ground, some principles regarding the structure of the problem. The situation here is analogous to that which leads to the feedback principle or the principle of optimality. Namely, given a process P, with a disturbance U and control M, the feedback principle states that the process should be controlled by observing the output and deriving the control so that the deviation of the output from a desired course is minimized. In a similar situation the optimality principle states that the control should be derived assuming that at the next decision time the control will again be selected on an optimal basis. The success in the application of these principles depends upon how much is known about the process, the disturbances, etc. At any rate, in order to proceed with control one has to start from some structural assumptions. Once a control principle is adopted, the control problem can be formulated and considered mathematically.

We shall approach the solution of the coordination problem by postulating two coordination principles.

Assume that the coordination object A_i is the interaction input object for the actual local processes, i.e., $A_i = U_i$ and $P_i : M_i \otimes U_i \rightarrow Y_i$. Furthermore, assume that B_i is void, i.e., the second level is not affecting the performance functions of the first level. The coordination process then proceeds as follows: The second level selects the intervention $\alpha_i \varepsilon U_i$ for each $i \varepsilon N$. The first-level units solve the problems $\{\Delta_i : i \varepsilon N\}$ by taking the process $P_{i\alpha} : M_i \otimes \{\alpha_i\} \rightarrow Y_i$. The derived control $\hat{m}_i(\alpha_i) \varepsilon M_i$ is then applied to the process.

The local control is shown as a function of the intervention α_i since for any α_i there is given a single m_i that represents the solution of the problem Δ_i. If the controls $\hat{m}_i(\alpha_i)$ are applied to the process $P : M \rightarrow Y$, the respective subprocesses $P_i : M_i \otimes U_i \rightarrow Y_i$ will be interconnected so that U_i comes from the outputs of other subsystems as specified by the interconnections. In particular, the input interaction to P_i will be $u_i(m(\alpha))$. We have then the following:

Principle of Interaction Prediction: The system is coordinable if for each $i \varepsilon N$ there exists $\hat{\alpha}_i \varepsilon A_i$ such that

$$\hat{\alpha}_i = u_i(\hat{m}(\hat{\alpha}))$$

The principle states that if the first-level units solve the local problems on the basis of the interactions which will actually occur after the control is applied, the overall problem is solved.

Using the interaction prediction principle, the coordination problem Δ^2 is reduced to the problem of predicting the interactions. If the second level can accurately predict the interactions, it can coordinate the first-level systems so that they solve the overall problem.

To formulate the second principle we shall assume that for each $i \varepsilon N$ the coordination object B_i is a set while A_i is a subset of the power set of U_i; i.e., the coordination consists in the specification of an element $\beta_i \varepsilon B_i$ (which determines the first-level performance G_i) and in the selection of a subset of interaction inputs $U_i' \subset U_i$. The subset U_i' can be thought of as representing the set of possible interactions. In other words, instead of trying to predict the interaction input, the second level predicts a set of interaction inputs which represent possible interactions for the local process. Since the first-level unit D_i has now the process $P_i : M_i \otimes U_i' \rightarrow Y_i$, it will solve the problem Δ_i by optimizing $G_i(\beta_i)$ with respect to both m_i and $\alpha_i \varepsilon U_i'$. Let $(\hat{m}_i(\beta_i), \hat{\alpha}_i(\beta_i)) \varepsilon M_i \otimes U_i'$

represent a pair that optimizes $G_i(\beta_i)$. Assume that for each $i \varepsilon N$, $\hat{m}_i(\beta_i)$ is applied to the process and the actual interaction input to P_i is $u_i(\beta)$, where $\beta = \{\beta_i : i \varepsilon N.\}$ We can formulate the following:

Principle of Interaction Balance: The system is coordinable if for each $i \varepsilon N$ there exists $\hat{\beta}_i \varepsilon B_i$ such that

$$u_i(\hat{\beta}) = \hat{\alpha}_i(\hat{\beta}_i)$$

The principle of interaction balance states that the overall problem is solved when the interactions which each of the first-level units require (in order to optimize local performance) are in agreement with the interactions which actually occur after the respective controls are applied.

The importance of coordination principles lies in the fact that they provide a basis for the structuring of multilevel systems in the sense that if one of the principles applies, the coordination problem can be formulated precisely. For example, assume that the conditions for the application of the interaction prediction principle exist. The coordination problem is then reduced to the correct prediction of interactions. The process evolves in the following way: The second level predicts the interactions and sends them down to the first-level units which then solve the local decision problems. If the interactions are correctly predicted, the solutions of the local problems imply the solution of the overall problem. If the prediction is not quite accurate, the difference between the predicted and actual interactions can be used as an indication of how closely the overall optimum is achieved. Furthermore, this difference can be used as a basis for an iterative procedure to solve a complex large-scale problem by decentralized, local, actions. Namely, on the basis of an initial guess of interactions by the second level, the first-level units can find the local solutions which result in certain values for actual interactions. On the basis of the difference between actual and predicted interactions, the second level makes an improved estimate of the interactions. The process proceeds iteratively until an equilibrium is reached.

The interaction balance principle can be similarly used. Namely, let each of the local units consider both the local control inputs and the interactions coming from other subsystems as independent variables and solve the local first-level problem by selecting the best values for both local input and interaction. If the interaction balance principle applies, such a procedure will yield the solution of the overall problem. If the interaction balance principle does not apply and provisions are made to modify the first-level performances, the second level can make these modifications on the basis of the discrepancy between the interactions as requested by the first-level units. Namely, the difference between the interactions required by a local unit and the interactions that have resulted from the action of other units indicates the discrepancy between the actual behavior of the overall system and the optimal behavior. The second level then modifies the first-level performances by selecting a new coordination term β, (e.g., in an iterative way) until the interactions are balanced. If the principle applies, this condition will imply the solution of the overall problem.

B. Goal Properties

The conditions for a successful application of the coordination principles can be given in terms of some rather general properties of systems. These properties refer to the existence or absence of conflict between the units on the same level as well as between the two levels.

1. Monotonicity Property:

For each $m \varepsilon M$ there is given a set of values for the performance functions, $v_1 = G_1(m), \cdots, v_n = G_n(m), v = G(m)$. There exists therefore a relation

$$\Psi \subset V_1 \otimes \cdots \otimes V_n \otimes V$$

such that $(v_1, \cdots v_n, v) \varepsilon \Psi$ if and only if there exists $m \varepsilon M$ such that $v_1 = G_1(m), \cdots, v = G(m)$. Often, Ψ is a function

$$\Psi : V_1 \otimes \cdots \otimes V_n \rightarrow V$$

We shall say that the family of performance functions $\{G, G_i : i \varepsilon N\}$ has the monotonicity property if Ψ is monotone in all its arguments, i.e.

$$G_1(m) \geq G_1(m') \wedge \cdots \wedge G_n(m) \geq G_n(m') \rightarrow G(m) \geq G(m')$$

Monotonicity eliminates certain conflicts between the units on two different levels. Namely, if the system has the goal monotonicity property, then the simultaneous improvements on the first level improve the performance on the second level.

2. Local Harmony:

Let $\hat{m} \varepsilon M$ be the optimal input for the second-level performance, i.e., for all m εM

$$G(\hat{m}) \leq G(m)$$

The system has local harmony if for all $i \varepsilon N$

$$G_i(\hat{m}_1, \cdots \hat{m}_{i-1}, m_i, \hat{m}_{i+1}, \cdots \hat{m}_n) \leq G_i(\hat{m}_1, \cdots \hat{m}_{i-1}, \hat{m}_i, \hat{m}_{i+1}, \cdots \hat{m}_n)$$

for any $m \varepsilon M$.

If the system has local harmony, then when the optimal controls for all other units other than D_i are given, the unit D_i itself will select the local control which is optimal for the second-level (overall) performance.

3. Global Harmony

Each of the first-level performance $G_i : i \varepsilon N$ depends upon the entire control input $m \varepsilon M$. Let $\hat{m} \varepsilon M$ be such that

$$G_i(\hat{m}^i) \leq G_i(m)$$

for all $m \varepsilon M$.

A system will have global harmony if for all $i, j \varepsilon N$, $\hat{m}^i = \hat{m}^i$.

Apparently, if the system has global harmony all decision units on the first level are in perfect agreement as far as the selection of the total control input is concerned. In this sense, the conflict between the units on the same level has been completely removed.

It is interesting to note that the local harmony and global harmony properties are not consecutively stronger. Namely, a system can have local harmony but not global harmony and vice versa. Investigations regarding the classes of systems which have monotonicity or harmony properties could be found in Mesarović *et al.* (1969).

C. Applicability of the Coordination Principles

The conditions for the applicability of the coordination principles can be expressed in terms of the goal properties introduced in the previous section. Detailed considerations of these relationships can be found in Mesarović *et al.* (1969). We shall mention here only some results of general interest.

Proposition 12. The principle of the interaction predictions applies whenever the system has local harmony.

Proposition 13. The principle of interaction balance applies whenever the system has both the monotonicity property and global harmony.

Proposition 14. Let M and Y be linear vector spaces, the processes P and $P_i : i \varepsilon N$, linear transformations, and $G_i : i \varepsilon N$, convex functions in all their arguments. Then the set of performances

$$G_i^* = G_i(m_i, y_i) + (\beta_i, u_i) - \sum_{j \neq i} (\beta_j, L(m_i, y_i))$$

where β_i and β_j are linear functionals, L is a linear operator and (,) denotes the inner product, satisfies the conditions of Proposition 14, i.e., the interaction balance principle applies.

The proofs of these propositions as well as further development in applications of the coordination principles can be found in Mesarović (1969). We shall present here only a simple example:

Let the overall process P be specified by the equations

$$y_1 = y_2 - m_1 - 2$$
$$y_2 = 2_{y_1} - m_2 + 1$$

and the overall performance function, G, be

$$v = m_1^2 + m_2^2 + y_1^2 + y_2^2$$

The overall problem is to find m_1 and m_2 so that v attains the minimum.

Assume that the system is decentralized so that there are two subunits on the first level. The first unit, D_1, has the process specified by

$$y_1 = u_1 - m_1 + 2$$

and the performance

$$v_1 = m_1^2 + y_1^2 + f_1(\alpha_1, \alpha_2)$$

where $f_1(\alpha_1, \alpha_2)$ represents the form of the second-level intervention and α_1, α_2 are the coordination terms.

Similarly, the second unit on the first level, D_2, is defined by

$$y_2 = 2u_2 - m_2 + 1$$
$$v_2 = m_2^2 + y_2^2 + f_2(\alpha_1, \alpha_2)$$

If D_1, after solving the local problem arrives at $\hat{m}_1$ which is optimal from the overall viewpoint and similarly D_2 arrives at the appropriate $\hat{m}_2$, the decentralization is successful. The problem for the coordinator is how to select α_1 and α_2 so that the decentralization is successful.

If the coordinator would like to apply the interaction balance principle, the first-level units have to be instructed to minimize their respective performance function with respect to both local control m_i as well as the interaction u_i. The structural problem then is how to select $f_1(\alpha_1, \alpha_2)$ and $f_2(\alpha_1, \alpha_2)$ so that there exist $\hat{\alpha}_1$ and $\hat{\alpha}_2$ such that the system has global harmony.

It is easy to see that the following two solutions are possible.

1. $f_i(\alpha_1, \alpha_2)$ is a linear function of all local output and the interactions

$$f_1(\alpha_1, \alpha_2) = \alpha_1 y_1 + \alpha_2 u_1$$
$$f_2(\alpha_1, \alpha_2) = -\alpha_2 y_2 - \alpha_1 u_2$$

Indeed, the system has global harmony for $\hat{\alpha}_1 = -\frac{2}{3}$ and $\hat{\alpha}_2 = \frac{8}{3}$; i.e., if the first-level performances are selected to be

$$v_1 = m_1^2 + y_1^2 - \tfrac{2}{3} y_1 + \tfrac{8}{3} u_1$$
$$v_2 = m_2^2 + y_2^2 + \tfrac{8}{3} y_2 - \tfrac{2}{3} u_2$$

The local optimal solution $\hat{m}_1 = -3$ and $\hat{m}_2 = 1$ will yield the overall optimal. The coordinator can arrive at the correct values for α_1 and α_2 by iteration.

2. Quite similarly, if the intervention function is quadratic

$$f_1(\alpha_1, \alpha_2) = \alpha_1 y_1^2 + \alpha_2 u_1^2$$
$$f_2(\alpha_1, \alpha_2) = \alpha_2 y_2^2 - \alpha_1 u_2^2$$

there exist values for α_1 and α_2 such that the system has global harmony, i.e., the overall solution is achieved by local optimization.

In Mesarović *et al.* (1969) it is shown that a quite analoguous decentralization can be applied to much more general situations. For example, similar procedures are developed for the linear dynamic systems with quadratic performance and arbitrary number of subsystems on the first level, for nonlinear systems and certain types of convex performance functions, etc. The details of these developments are outside of the scope of this paper. These further results are of interest here only as an illustration of the application of the general systems study of decentralization. The coordination principles, the harmony properties, and the relations between the two, as investigated on the general systems level, provide the basis for the structuring of decentralized systems which can be then used for the solution of the large organizational problems in the more specific framework. The two coordination principles proposed here appear to have considerable application. However, they are certainly not the only coordination principles. It is therefore equally important that the discussion in this chapter indicate an approach of how the general systems theory can be applied for the study of decentralization problems. Using general systems theory one can investigate on a mathematical basis some other coordination principles.

D. Further Comments

The structural problems of coordination considered in this section represent only one aspect of a more general study of multilevel systems. The procedural problem (how to find a solution for Δ^2) has also been investigated in some detail (Mesarović, 1969). Furthermore, the application of the theory of multilevel systems in human organizations has been considered (Mesarović *et al.* 1964). A summary of these developments will be published shortly in Mesarović (1969).

ACKNOWLEDGMENT

Research reported in this paper has been supported in part by ONR Contract 1141(12).

REFERENCES

Mesarović, M. D. 1963. "A Unified Theory of Learning and Information," I Computer and Information Sciences Symposium, Spartan Press, Washington, D.C.

Rasiowa, H. and R. Sikorski, 1963. *Mathematics of Metamathematics,* Polish Academy of Science, Math. Monograph 41, Warsaw.

Mesarović, N. D. 1964a. "Toward a Formal Theory of Problem Solving," Symposium on Computer Augmentation of Human Reasoning, Spartan Press, Washington, D.C.

Mesarović, M. D. 1964b. "Foundations for a General Systems Theory," in *Views on General Systems Theory,* John Wiley & Sons, New York.

Mesarović, M. D., J. L. Sanders, and C. F. Sprague. 1964. *An Axiomatic Approach to Organizations from a General Systems Viewpoint,* John Wiley & Sons, New York.

Mesarović, M. D. 1966. "On the Auxiliary Functions and Constructive Specification of the General Time Systems," SRC Report 85-A-66-33, Case Institute of Technology, Cleveland, Ohio.

Mesarović, M. D. 1967a. "General Systems Theory and Its Mathematical Foundations," 1967 Systems Science and Cybernetics Conference, IEEE, Boston, Massachusetts.

Mesarović, M. D. 1967b. "Mathematical Theory of General Systems and Some Economic Problems," *Proc. of 1967 Varenna Conference on Mathematical Systems Theory and Economics*, Springer-Verlag, Berlin.

Mesarović, M. D. 1968a, "Auxiliary Functions and Constructive Specifications of General Systems," *Math. Systems Theory J.*, No. 2.

Mesarović, M.D. 1968b. "Systems Theory and Biology—View of a Theoretician," in *Proceedings of the Third Systems Symposium*, Springer-Verlag, Berlin.

Mesarović, M. D., D. Macko, and Y. Takahara. 1969. *Theory of Multi-level Systems*, Academic Press, New York.

Windeknecht, T. G. 1967. "Mathematical Systems Theory: Causality," *Math. Systems Theory. J.* Vol. 2 pp. 279–288.

Windeknecht, T. G. and M. D. Mesarović. 1967. "On General Dynamical Systems and Finite Stability," in *Differential Equations and Dynamical Systems*, Academic Press, New York.

3

A SYSTEMS-THEORETIC APPROACH TO DECOMPOSITION OF TWO-PERSON BOARD GAMES

RANAN B. BANERJI

I. INTRODUCTION

Researches in general systems theory and artificial intelligence at the Case Western Reserve University are heavily concentrated in the Systems Research Center and the two research activities have greatly influenced one another. One of the basic beliefs that motivates activities at the Center is that many seemingly dissimilar phenomena may be studied with similar techniques at a suitable level of abstraction. The abstraction required to unify dissimilar phenomena necessarily ignores certain detailed structures of the problem areas involved: however, often enough structure is left to enable the application of powerful techniques of analysis.

Such unified treatment of seemingly diverse phenomena is certainly not new in the field of science. The "essential similarity" between electrostatic fields and steady-state flow of perfect incompressible fluids has been known for years; so have the similarities between the velocities of point masses and currents in lumped-parameter electrical networks. It has been known for a long time that integers under addition and one-one maps of any set into itself under composition have enough essential similarities to make it meaningful to talk about cosets, normal subgroups, and quotient groups in both cases. Nor does meaningful abstraction cease there. The "similarity" between the homomorphisms from a group onto a quotient group and the homomorphism from a ring onto a quotient ring turn out to be special cases of homomorphisms of algebras onto a quotient algebra. The fact is that the kernels of such homomorphisms are congruences—a phenomenon which goes beyond algebras with an underlying group structure.

There is thus a considerable body of evidence which indicates the feasibility of unified treatment of diverse phenomena. Another small case history was added to this body at the Systems Research Center when Koffman (1967), abstracting on a class of two-person board games, unified them to sufficiently meaningful structure such that a single learning program could improve its performance at playing any game in this class. The present paper, it is believed, adds another such case history. In what follows, a method for simplifying solutions of a wide class of board games will be discussed. Since board games under suitable abstraction are similar to automata, the techniques of parallel decomposition of automata are equally applicable to board games. The method of simplification to be discussed here consists of a specialization of parallel decomposition. The technique is useful because of certain properties of graphs which (again at a suitable level of abstraction) are similar to automata.

In what follows, a general system under control and disturbance (an "M-system") will be introduced and certain facts about it stated without proof. Board games will then be introduced and the isomorphism between board games and M-systems indicated. This will establish the existence of uniform solution methods for all board games. Board games will then be specialized to a class called "graph-interpretable" and their similarity to automata and graphs will be evident. Certain facts regarding decomposition of graphs will then be quoted without proof. A test for the decomposability of a graph-interpretable game will then be introduced and justified. A simple example will then be given of the application of the test and the use of the resulting decomposition.

II. SOLUTION OF BOARD GAMES AND CONTROL OF SYSTEMS

The model of a system under control and disturbance that will be called an "M-situation" below was discussed by Marino (1966).

An M-situation is a 6-tuple $\langle S,C,D,M,S_W,S_L\rangle$ where S,C, and D are sets (of "states," "controls," and "disturbances" respectively), S_W and S_L are disjoint subsets of S and M is a map from a subset of $SXCXD$ onto S. M has the property that if $M(s,c',d)$ is defined for some d and

$M(s,c,d')$ is defined for some c, then $M(s,c,d)$ is defined.

For each $c \in C$ and $d \in D$ define

$S_c = \{s \mid (Es')(Ed)[M(s,c,d)=s']\}$

$S_d = \{s \mid (Es')(Ec)[M(s,c,d)=s']\}$

Also denote $M(s,c,d)$ by $(c,d)(s)$

A control strategy for an M-situation is a map

$$P_C : \bigcup_{c \in C} S_c \rightarrow C$$

such that

$$P_C(s)=c \text{ implies } s \in S_c$$

A disturbance strategy for an M situation is a map

$$P_D : \bigcup_{d \in D} S_d \rightarrow D$$

such that

$$P_D(s)=d \text{ implies } s \in S_d$$

Given a state

$$s_0 \in \bigcup_{c \in C} S_c - S_W - S_L$$

a control strategy P_C is called a *winning strategy* for s_0 if there exists a finite sequence (c_i,d_i) $(1 \leqslant i \leqslant n)$ such that

$$c_1 = P_C(s_0),\ d_1 = P_D(s_0);$$

and for each i $(1 \leqslant i < n)$:

$$c_{i+1} = P_C((c_i,d_i)((c_{i-1}d_{i-1})(\ldots(c_1,d_1)(s_0)\ldots),$$
$$d_{i+1} = P_D((c_i,d_i)((c_{i-1},d_{i-1})(\ldots(c_1,d_1)(s_0\ldots);$$

and

$$(c_n,d_n)((c_{n-1},d_{n-1})(\ldots(c_1,d_1)(s_0)\ldots) \in S_W.$$

One can express the above by saying that a winning strategy is one such that, no matter what strategy is chosen by the disturbing influence, any sequence of applications of controls and disturbances applied according to the strategies results in a winning situation.

A control strategy P_C is called a *nonlosing strategy* for s_0 if it is either a winning strategy or for no disturbance strategy P_D is it the case that there exists a sequence (c_i,d_i) ($1 \leqslant i \leqslant n$ as before) such that

$$c_1 = P_C(s_0),\ d_1 = P_D(s_0);$$

and for each i $(1 \leqslant i < n)$

$$c_{i+1} = P_C((c_i,d_i)((c_{i-1},d_{i-1})(\ldots(c_1,d_1)(s_0)\ldots),$$
$$d_{i+1} = P_D((c_i,d_i)((c_{i-1},d_{i-1})(\ldots(c_1,d_1)(s_0)\ldots);$$

and

$$(c_n d_n)((c_{n-1},d_{n-1})(\ldots(c_1 d_1)(s_0)\ldots) \in S_L$$

A situation for which a winning strategy exists is called a *forcing situation.* The set of all forcing situations is denoted by S_F. A situation for which a nonlosing strategy exists but no winning strategy exists is called a *neutral situation.*

The following theorem is of great interest. We shall state it here without proof (a proof can be found in Marino, 1966).

Theorem 2.1. Given an M-situation, there exists a strategy which is a winning strategy for every forcing situation and a nonlosing strategy for every neutral situation.

A game situation is an M-situation with two specified elements $c_0 \in C$ and $d_0 \subset D$ (the "inactions") such that

(G1) $S_c \cap S_d \neq \Phi$ implies either $c=c_0$ or $d=d_0$ but not both

(G2) $S_W \cup S_L \subseteq S - \bigcup_{c \in C} S_c$

(G3) $s \in S_{c_0} \cap S_d$ implies $(c_0,d)(s) \in S - S_W - S_{c_0}$

(G4) $s \in S_c \cap S_{d_0}$ implies $(c,d_0)(s) \in S - S_L - S_{d_0}$

A board game is a 5-tuple $\langle S,G,F,W,L \rangle$ where S is a set, F and G are sets of functions each mapping subsets of S into S (i.e., $t \in F \cup G$ implies t: $S_t \rightarrow S$ where $S_t \subseteq S$), and W and L subsets of S such that

(B1) $(\bigcup_{f \in F} S_f \cap (\bigcup_{g \in G} S_g) = \emptyset$

(B2) $W \cap L = \Phi$

(B3) $W \cup L \subseteq S - \bigcup_{f \in F} S_f - \bigcup_{g \in G} S_g$

(B4) $s \in S_f$ and $f \in F$ implies $f(s) \in S - \bigcup_{f \in F} S_f - L$

(B5) $s \in S_g$ and $g \in G$ implies $g(s) \in S - \bigcup_{g \in G} S_g - W$

In effect, the axioms say the following, "The players play alternately. The game stops whenever a win or loss is reached. The opponent cannot make a final move into a loss and the player cannot make a final move into a loss."

A function Q_F: $\bigcup_{f \in F} S_f \rightarrow F$ is called a *board control strategy* if for all s, $Q_F(s)=f$ implies $s \in S_f$. Similarly, a function Q_G: $\bigcup_{g \in G} S_g \rightarrow G$ is called a *board disturbance* strategy if for all s, $Q_G(s)=g$ implies $s \in S_g$.

A board control strategy Q is called *winning* for $s_0 \in S$ if there exists an integer N such that for every board disturbance strategy Q_G there exist sequences $(f_1,f_2,\cdots,f_n; f_i \in F$ for all i, $1 \leqslant i \leqslant n)$ and $(g_1,g_2,\cdots,g_{n-1}; g_i \in G$ for all i, $1 \leqslant i \leqslant n)$ such that

(a) $n \leqslant N$

(b) $Q_F(s_0)=f_1; Q_G(f_1(s_0))=g_1$

$Q_F(g_1(f_1(s_0)))=f_2;\ Q_G(f_2(f_1(s_0))))=g_2$

for all $i<n-1$

$Q_F(g_i(f_i \cdots (f_1 s_0) \cdots)=f_{i+1}$

$Q_G(f_{i+1}(g_i \cdots f_1(s_0 \cdots) = g_{i+1}$
$Q_F(g_{n-1}(f_{n-1} \cdots g_1(f_1(s_0) \cdots) = f_n$

and

(c) $f_n(g_{n-1}(f_{n-1} \cdots g_1(f_1(s_0) \cdots) \in W$

The following theorem will be stated without proof (the proof can be found in Banerji, 1967)

Theorem 2.2. There is a one-one mapping B from the set of all game situations onto the set of all board games and for each M situation R a one-one mapping T_R from the set of all control strategies of R to all board control strategies of $B(R)$ and a one-one map K_R from the states of R onto the states of $B(R)$ such that $P_{C'}$ is winning for s_0 in R if and only if $T_R(P_C)$ is winning for $K_R(s_0)$ in $B(R)$.

This theorem, together with Theorem 2.1 establishes the existence of at least one board control strategy winning for every state in a board game from which a win is possible.

In what follows, a method for finding winning strategies in a class of board games will be discussed.

III. GRAPH-INTERPRETABLE GAMES AND SUMS OF GRAPHS

A board game $\langle S,F,G,W,L \rangle$ will be called *graph interpretable* if and only if there exists an abstract set Ω, a set H of functions mapping subsets of Ω into Ω (i.e., $h \in H$ implies $h: S_h \rightarrow \Omega$ where $S_h \subseteq \Omega$) and a subset T of Ω such that

(Gr1) $S = \Omega \otimes \{0,1\}$

(Gr2) $W = \{(s,1) \mid s \in T\}$
$L = \{(s,0) \mid s \in T\}$

(Gr3) $h \in H$ if and only if (1) there exists an $f \in F$ such that $s \in S_h$ if and only if $(s,0) \in S_f$ and $f(s,0) = (h(s),1)$
(2) there exists a $g \in G$ such that $s \in S_h$ if and only if $(s,1) \in S_g$ and $g(s,1) = (h(s),0)$

(Gr4) $(s,k) \in S_f$ and $f \in F$ implies $k = 0$
$(s,k) \in S_g$ and $g \in G$ implies $k = 1$

Theorem 3.1. A board game is graph interpretable if and only if there are two subsets S_0,S_1 of S and two one-one onto maps

$\alpha: S_0 \rightarrow S_1$
$\beta: F \rightarrow G$

such that

(1) $L\subseteq S_0$; $W\subseteq S_1$

(2) $\bigcup_{g\in G} S_g\subseteq S_1$; $\bigcup_{f\in F} S_f\subseteq S_0$

(3) $\alpha^{-1}(S_g)=S_{\beta-1(g)}$; $\alpha(S_f)=S_{\beta(f)}$

(4) $\alpha^{-1}(f(s))=\beta(f)(\alpha(s))$; $\alpha(g(s))=\beta^{-1}(g)\ (\alpha^{-1}(s))$

for each $f\epsilon F$, $g\epsilon G$ and any $s\epsilon S$ for which either side of the equation is defined.

(5) $\alpha(L)=W$

(6) $S_0\cup S_1=S$

Proof. Let $\langle S,F,G,W,L\rangle$ be graph interpretable. Then there exists a set Ω, a set H of partial functions mapping subsets of Ω into Ω, and a subset T of Ω satisfying Gr1 to Gr4. Define $S_0=\{(s,0)\mid s\epsilon\Omega\}$ and $S_1=\{(s,1)\mid s\varepsilon\Omega\}$. Then (6) is satisfied.

Define $\alpha(s,0)=(s,1)$ for each element of S_0. This is one-one from S_0 onto S_1. Also $(s,k)\epsilon S_g$ for $g\epsilon G$ implies $k=1$ whence $(s,k)\epsilon S_1$. Hence $\bigcup_{g\epsilon S_g} S_g\subseteq S_1$. Similarly $\bigcup_{f\epsilon F} S_f\subseteq S_0$ satisfying (2).

$$W=\{(s,1)\mid s\epsilon T\}\subseteq\{(s,1)\mid s\varepsilon\Omega\}=S_1$$

Similarly, $L\subseteq S_0$ proving (1).

If $(s,0)\epsilon L$ then $s\epsilon T$ and $\alpha(s,0)=(s,1)$ whence $\alpha(s,0)\epsilon\{(s,1)\mid s\epsilon T\}=W$. Hence $\alpha(L)\subseteq W$. It can be shown similarly that $\alpha^{-1}(W)\subseteq L$ proving (5).

For each $f\epsilon F$ there exists an $h\epsilon H$ and $g\epsilon G$ such that $S_f=\{(s,0\mid s\epsilon S_h\}$ and $S_g=\{(s,1)\mid s\epsilon S_h\}$, $f(s,0)=(h(s),1)$ and $g(s,1)=(h(s),0)$. If one defines $\beta(f)=g$, the resulting β is one-one onto. Also (3) and (4) will be satisfied by this β.

Let now $\langle S,F,G,W,L\rangle$, S_0, S_1, α and β be as defined by (1) through (6). Then one can define Ω, H and T as follows: $\Omega\subseteq S\otimes S$ such that $(s_0,s_1)\epsilon\Omega$ if and only if $s_1=\alpha(s_0)$. Hence $s_0\epsilon S_0$ and $s_1\epsilon S_1$. Denote s_0 by $\{(s_0,s_1),0\}$ and s_1 by $\{(s_0,s_1),1\}$. Since α is one-one onto, and because of (6), if $s\epsilon S$ then either $s\epsilon S_1$ and $s=\{(\alpha^{-1}(s),s),1\}$ or $s\epsilon S_0$ and $s=\{(s,\alpha(s)),0\}$. Hence $S=\Omega\otimes\{0,1\}$ satisfying Gr1.

$s\epsilon L\subseteq S_0$ if and only if $\alpha(s)\epsilon W\subseteq S_1$. Define by T the set of all pairs $(s,\alpha(s))$ such that $s\epsilon L$. Then $s\epsilon T$ if and only if $(s,0)\epsilon L$ and $(s,1)\epsilon W$. This establishes Gr2.

$H\subseteq\{h\mid h:S_h\to\Omega\}$ is constructed as follows. Let $f\epsilon F$ and $s\epsilon S_f$. Then by (2) and the construction of Ω, $s=\{(s,\alpha(s)),0\}$. By (3) $\alpha(s)\epsilon S_{\beta(f)}$. Define h so that

$$S_h=\{(s,\alpha(s))\mid s\epsilon S_f\}=\{(s,\alpha(s))\mid\alpha(s)\epsilon S_g\}$$

and $h(s,\alpha(s))=(\beta(f)(\alpha(s)),\ f(s))$; $h(s,\alpha(s))\epsilon\Omega$ since by (4) above $\alpha(\beta(f)(\alpha(s)))=f(s)$. Also, $f(s)=(h(s,\alpha(s)),1)$. This establishes the first part of Gr3 and Gr4. The second part follows similarly.

This theorem is included to show clearly what kind of symmetry is demanded of a graph-representable game. The definition, based on the existence of the graph, did not clarify the structure sufficiently.

A graph-interpretable game is isomorphic to what Berge has called a "NIM-type" game (Berge, 1962). In these games a directed graph is given and a counter is placed on a specified node of the graph. Each player in his turn moves the counter from its present node to an adjacent node along an arc. The player who cannot make a move loses.

A graph-interpretable game is completely defined by the triple $\langle \Omega, H, T \rangle$. For every graph-interpretable game one can define the graph $\langle \Omega, \bigcup_{h \in H} h\}$.

A graph $\langle \Omega, R \rangle$ where $R \subseteq \Omega \otimes \Omega$ is called *progressively bounded* if the length of all chains in $\langle \Omega, R \rangle$ is bounded. A graph-interpretable game is called *Grundy-tractable* if the corresponding graph is progressively finite and if $W \cup L = S - \bigcup_{f \in F} S_f - \bigcup_{g \in G} S_g$ i.e., if in the corresponding triple $\langle \Omega, H, T \rangle$, $s \notin \bigcup_{h \in H} S_h$ if and only if $s \in T$ i.e., $T = \{s \mid s \in \Omega$ and $R(s) = \Phi\}$ in the corresponding graph $\langle \Omega, R \rangle$.

Given a progressively bounded graph $\langle \Omega, R \rangle$ one can define the following integral-valued function M on its nodes, called the Grundy function.

$$M(s) \geqslant 0 \text{ for all } s \in \Omega$$

$M(s) = n$ implies $M(s') \neq n$ for all $s' \in R(s)$ and for each $m < n$ there exists an $s' \in R(s)$ such that $M(s') = m$.

The importance of a Grundy function to Grundy-tractable games is brought out by the following definitions and well-known theorems.

Given a board game, let S' be the set of states for which a winning strategy exists. A board control strategy is called *cautious* if and only if $s \in S'$ implies that either $Q_F(s)(s) \in W$ or for every $g \in G$ such that $Q_F(s)(s) \in S_g$, $g(Q_F(s)(s)) \in S'$.

A stronger form of the following theorem is proved in Banerji (1967).

Theorem 3.2. In a Grundy-tractable game a cautious strategy is a winning strategy for every element of S'.

The relationship between the Grundy function and the set S' is given by the following theorem which is a direct interpretation of a theorem in Berge (1962).

Theorem 3.3. In a Grundy-tractable game

$$S' = \{(s,0) \mid M(s) > 0\}$$

If the graph corresponding to a game has a small number of nodes, then the value of the Grundy function of each of its nodes can be calculated by an exhaustive procedure. This, of course, is impossible in any realistic game where the nodes run into superastronomical numbers. However, many graphs can be decomposed into "sums" of smaller graphs, whose Grundy functions can be calculated individually and from these the Grundy function of the original graph can be calculated by an extremely straightforward process.

Given a finite set of graph-interpretable games $\{\langle \Omega_i, H_i, T_i \rangle \mid 1 \leqslant i \leqslant n\}$, a graph-interpretable game $\{\Omega, H, T\rangle$ is called the sum of $\{\Omega_i, H_i, T_i\rangle\}$ if and only if

(S1) $\Omega = \Omega_1 \otimes \Omega_2 \otimes \cdots \otimes \Omega_n$

(S2) $h \in H$ and $h(s_1, \cdots, s_n) = (s_1', \cdots, s_n')$ if and only if there is a unique positive integer $i \leqslant n$ and member $h_i \in H_i$ such that
$s_j' = s_j$ if $j \neq i$
$s_i' = h_i(s_i)$

(S3) $T = T_1 \otimes T_2 \otimes \cdots \otimes T_n$

To indicate the method of calculation, one needs to define a special binary operation $\oplus$ between nonnegative integers. Let a and b be two such integers. Let

$$a = a_0 + a_1 2 + a_2 2^2 + \cdots + a_m 2^m \; 0 \leqslant a_i \leqslant 1$$
$$b = b_0 + b_1 2 + b_2 2^2 + \cdots + b_n 2^n 0 \; \leqslant b_i \leqslant 1$$

i.e., let $a_m a_{m-1} \cdots a_0$ and $b_n b_{n-1} \cdots + b_0$ be the binary representation a and b. One can assume without loss of generality that $m = n$ and that some of the leading binary digits are 0.

One defines

$$c = (a \oplus b) = c_0 + c_1 2 + c_2 2^2 + \cdots + c_m 2^m$$

where for each i, $c_i = a_i + b_i$ (mod. 2).

It can be seen easily that the $\oplus$ operation is a group operation on integers, with 0 as the unit element and every integer its own inverse.

The following theorem indicates the use of the $\oplus$ operator in the calculation of the Grundy functions of sums of games.

Theorem 3.3a. (Berge) Let $\{\langle \Omega_i, H_i, T_i \rangle \mid i = 1, 2, \cdots n\}$ be a collection of graph-interpretable Grundy-tractable games and let M_i be their Grundy functions. One defines M on their sum as follows

$$M(s_1, \cdots, s_n)) = M_1(s_1) \oplus M_2(s_2) \oplus \cdots \oplus M_n(s_n).$$

M is a Grundy function on the sum.

Theorem 3.4. A graph-representable game $\langle \Omega, H, T \rangle$ is the sum of a set of n graph-representable games $\{\langle \Omega_i, H_i, T_i \rangle \mid 1 \leqslant i \leqslant n\}$ if and only if there exists a set of n equivalence relations $E_i (1 \leqslant i \leqslant n)$ on Ω and a set of disjoint subsets H_i' $(1 \leqslant i \leqslant n)$ of H such that

(1) $\cup H_i' = H$

(2) $\cap E_i = I$, the identity relation on Ω

(3) For any $s_1s_2\cdots s_n(s_i\in\Omega)$ there exists $s\in\Omega$ such that s_iE_is for each i.
(4) $s\notin T$ implies that for some E_i such that sE_is' implies $s'\notin T$
(5) sE_is' implies $s\in S_h$ if and only if $s'\in S_h$ for all $h\in H_i$ and $h(s)E_ih(s')$ for all h such that $s\in S_h$ and $s'\in S_h$.
(6) $h\in H_i'$ implies for all $s\in S_h$, $h(s)E_js$ for all $j\neq i$.

Proof. The sufficiency is proved by constructing $\{\langle\Omega_i,H_i,T_i\rangle\}$ as follows: Ω_i is the set of equivalence classes of E_i. Since $E_i=I$, two distinct elements of Ω do not lie in the same equivalence class of every E_i. Hence $\Omega\subseteq\Omega_1\otimes\Omega_2\otimes\cdots\otimes\Omega_n$. Let $s\in\Omega_1\otimes\Omega_2\otimes\cdots\otimes\Omega_n$. If $s\notin\Omega$, then there is some set of equivalence classes $e_1,e_2,\cdots,e_n(e_i$ an equivalence class of E_i for each i such that $e_1\cap e_2\cdots\cap e_n=\Phi$. Take an element $s_1\in e_1$, $s_2\in e_2,\cdots,s_n\in e_n$. Then there exists (by 3 above) an $s\in e_1\cap e_2\cdots\cap e_n$. This contradicts $e_1\cap e_2\cdots\cap e_n=\Phi$. Hence $s\in\Omega$ proves $\Omega_1\otimes\Omega_2\otimes\cdots\otimes\Omega_n\subseteq\Omega$. S1 is thus established.

Let T_i be the set of equivalence of E_i which contains some element $s\in T$. Clearly, then $T\subseteq T_1\otimes T_2\otimes\cdots\otimes T_n$, again since $\cap E_i=I$. Let $s\notin T$; for each i let e_i be the equivalence class of E_i containing s. By (4), there exists some E_i such that $s'\notin T$ for no member s' of e_i. Hence $e_i\notin T_i$. Hence $s\notin T_1\otimes T_2\otimes\cdots\otimes T_n$, hence the complement of T is contained in the complement of $T_1\otimes T_2\otimes\cdots\otimes T_n$ or $T_1\otimes T_2\otimes\cdots T_n\subseteq T$. This establishes S3.

Define H_i as follows. For every $h'\in H_i'$, define a member $h\in H_i$, such that if $s\in S_{h'}$, then the equivalence class e_i of E_i containing s is a member of S_h and $h(e_i)$ is the equivalence class of E_i containing $h'(s)$. By (5), this determines the function h unequivocally. For all $E_j(j\neq i)$, if e_j is the equivalence class of E_j containing s, then by (6) e_j is also the equivalence class containing $h'(s)$. Hence if $h'\in H_i'$, then the $h'(s)$ is in the intersection of the blocks of $E_j(j\neq i)$ containing s and the block $h'(e_i)$. This establishes S2, indicating that conditions (1–6) are sufficient for (S1–S3).

To show necessity let $\langle\Omega,H,T\rangle$ be the sum of $\{\langle\Omega_i,H_i,T_i\rangle\,|\,(1\leqslant i\leqslant n\}$. Define $(s_1,s_2,\cdots,s_n)$ $E_i(s_1',s_2'\cdots s_n')$ if and only if $s_i=s_i'$. Clearly $(s_1,s_2,\cdots,s_n)(\cap E_i)(s_1',s_2'\cdots s_n')$ if and only if $s_i=s_1'$ for every i. This establishes (2).

Since $T=T_1\otimes T_2\otimes\cdots\otimes T_n$, by (S3) $(s_1,s_2,\cdots,s_n)\notin T$ implies $s_i\notin T_i$ for some i. Hence for any $(s_1',s_2'\cdots s_n')$ $s_i=s_i'$ implies $(s_1',s_2',\cdots,s_n')\notin T_i$. Hence (4) follows.

Now for every $h\in H$ there is an unique integer $i(1\leqslant i\leqslant n)$ such that $h(s_1,\cdots,s_n)=(s_1',s_2',\cdots,s_n')$ implies $s_i\in S_{h'}$ for some $h'\in H_i$ $s_i'=h'(s_i)$ and $s_j'=s_j$ for all $j\neq i$. Define the class of subsets $\{H_i'\}$ as follows: $h\in H_i'$ if and only if the corresponding $h'\in H_i$. Since there is an unique i with this property for every element of H, the subsets H_i' are disjoint. Since an i exists for every element $h\in H$, (1) follows. Also, if $(s_1,s_2,\cdots,s_n)$

$E_i(s_1', s_2', \cdots, s_n')$, then $s_i = s_i'$. Hence if $(s_1, s_2, \cdots, s_n) \epsilon S_h$ for $h \epsilon H_i'$ then $s_i \epsilon S_h'$, hence $(s_1', s_2' \cdots, s_n') \epsilon S_h$ also. Again $h((s_1, s_2, \cdots, s_n)) = (s_1, s_2, \cdots, s_{i-1}, h'(s_i), s_{i+1}, \cdots, s_n)$ and $h((s_1', \cdots, s_n')) = (s_1', s_2', \cdots, s_{i-1}', h'(s_i), s_{i+1}, \cdots, s_n)$ whence $h((s_1, s_2, \cdots, s_n) E_i h(s_1', s_1', \cdots, s_n')$. Also, $h((s_1, s_2, \cdots, s_n)) = (s_1, s_2, \cdots, h(s_i) \cdots s_n)$ so that $h((s_1, s_2, \cdots, s_n)) E_j (s_1, s_2, \cdots, s_n)$ for all $j \neq i$. This establishes (5) and (6).

Let there be n elements $s_1, s_2, \cdots, s_n$ in Ω. Denote these by $(s_{11}, \cdots, s_{1n})$, $(s_{21}, \cdots, s_{2n}) \cdots (s_{n1} \cdots s_{nn})$ respectively. Let $s' = (s_{11}, s_{22}, \cdots, s_{nn})$. Then $s_i E_i s_i$ for each i. This establishes (3).

IV. AN EXAMPLE

The procedures discussed in the last section will be exemplified by pointing out how they lead readily to the well-known procedure for winning in the game of NIM (Gardner, 1958). The game can be formalized as $\langle \Omega, H, T \rangle$ where Ω is the set of triples (χ_1, χ_2, χ_3) where $\chi_1 \leqslant 3, \chi_2 \leqslant 5$ and $\chi_3 \leqslant 7$, $(\chi_1', \chi_2', \chi_3')$ $\overset{\cup}{h \epsilon} Hh((\chi_1, \chi_2, \chi_3))$ if and only if for some j $(1 \leqslant j \leqslant 3)$ $\chi_j \langle \chi_j'$ and $\chi_i' = \chi_i$ for all $i \neq j$. T consists of the signal node (0,0,0). One defines an equivalence relation on the members of Ω by

$$(\chi_1, \chi_2, \chi_3) E_i (\chi_1', \chi_2', \chi_3')$$

if and only if $\chi_i = \chi_i'$

and a partition on H by

$$h \epsilon H_i \text{ if and only if}$$
$$h(\chi_1, \chi_2, \chi_3) = (\chi_1', \chi_2', \chi_3')$$

implies $\chi_i \neq \chi_i'$.

It can be seen readily that the conditions of Theorem 3.4 are satisfied. Hence one obtains three Grundy-tractable games, $\{\langle \Omega_i, H_i, T_i \rangle \mid i = 1,2,3\}$. Ω_i consists of the integers χ_i and $h(\chi_i) < \chi_i$ for each $h \epsilon H_i$. It is readily seen that the Grundy function of each game is defined by $M_i(\chi_i) = \chi_i$, since $M_i(0)$ is evidently 0 and χ_i can be reduced to any integer χ_i' $(0 < \chi_i' < \chi_i)$. Thus if $\chi_1 \oplus \chi_2 \oplus \chi_3 > 0$, then the player whose move it is wins by reducing the proper χ_i to reduce the sum to 0.

ACKNOWLEDGMENTS

The research reported in this paper was supported by the U.S. Air Force Office of Scientific Research under Grant No. 125–67 and by the National Science Foundation under Grant No. GK 1386.

REFERENCES

Banerji, R. 1967. "An Approach to a Theory of Problem Solving," Systems Research Center, Case Institute of Technology, Cleveland, Ohio.

Berge, C., 1962. *The Theory of Graphs and Its Application,* John Wiley & Sons, New York.

Gardner, M., 1958. "Nim and Variations," *Scientific American* **198** (1): 104.

Koffman, E. "Learning Through Pattern Recognition Applied to a Class of Games," Systems Research Center Report SRC 107-A-67-45, Case Institute of Technology, Cleveland, Ohio.

Marino, L. 1966. "Winning and Non-Losing Strategies in Games and Control," Systems Research Center Report SRC 91–A-66-36, Case Institute of Technology, Cleveland, Ohio.

4

CONTINUITY

PRESTON C. HAMMER

I. GENERALIZATION OF CONTINUITY

As a technical word, appearing in the calculus, continuity requires infinite limiting processes or the equivalent ε, δ treatment of Peano. It would appear that Cauchy introduced the word in connection with analysis and it would be safe to conjecture that he had a reason for selecting such a word from the common language. That reason might be based on graphs of functions which appear disconnected at a point of discontinuity and there was and is a natural association of arcwise connectedness and continuity. However that may be, continuity is not best seen from that standpoint.

Later on, the analysts and topologists broke the first barrier in the discussion of continuity via neighborhoods. This new continuity of functions or mappings can be depicted as follows. Let t map a set M into another set M_1. Let p be a point of M and let $tp=q$ be the corresponding point in M_1. Then t is continuous at p, provided that to every neighborhood V of q there is a neighborhood U of p, the image of which is contained in V. Since a neighborhood V of q contains all the points which may be considered close to q, this technical statement, in effect, says that t is continuous provided points close to p map into points close to q. Then why not say it that way? Here is the clue to the generalization I now propose.

A mapping from one set to another is continuous provided it preserves closeness. In topology a set is "close" to a point provided the point is in the set or is a limit point of the set. The technical way of writing this is now important. Let fX be the closure of a set X contained in M. That is, fX is the union of X and all limit points of X.

Now let gY similarly be the closure of a set Y contained in M_1. Then t is continuous at p provided $p \epsilon fX$ implies $tp \epsilon g(tX)$. That is, if t is continuous at p and X is close to p, then the image tX of X is close to the image tp of p. Dropping part of the paraphernalia, we have "if X is close to p then tX is close to tp." The following statements are equivalent:

(1) t: $M \rightarrow M_1$ is continuous at p for each $p \epsilon M$
(2) $p \epsilon fX \rightarrow tp \epsilon g(tX)$, for each $p \epsilon M$, $X \subseteq M$
(3) $t(fX) \subseteq g(tX)$, for each $X \subseteq M$
(4) $tf \subseteq gt$

The last forms (3, 4) display only global continuity. Form 4 corresponds to one verbal way of defining t to be continuous at every point; i.e., the transform of the closure is contained in the closure of the transform.

So far any deviation from standard practice in topology has not been suggested. The main deviation is the use of a better notation for closures as functions mapping sets into sets. This, as we shall see, is a significant deviation. Now comes the second lession in generalization. The topologists provided the first one and I have rewritten their definition in a better form for one purpose—generalizations.

Generalization is achieved by weakening or removing conditions, it is suggested by curiosity, by suspicion, by forms, or by inadequacy of a given concept for applications. For example, the use of "closeness preserving" might make one wonder why topologists chose such a strange form of "closeness." This would lead to the realization that the topologists actually had one model in mind, the real continuum and its generated Euclidean spaces but not approximation and certainly not finite sets. Thus there is no reason to accept the properties of the functions f and g prescribed by topologists. The form of the definition in parts 2, 3, or 4 suggests that a generalization is at hand simply by using the same form but not requiring f and g to be topological closures.

How far dare we go with this idea? Strangely enough, there is often little courage or imagination displayed in mathematics. Certainly to make trial explorations of this notion would not seem to require much courage. The difficulty is that mathematics is an autocratic field and that, naturally, few mathematicians dare challenge the definitions of concepts once they are widely accepted. Yet for the purposes of computing and systems theory, we need something like continuity, but it must apply to finite sets of objects.

Let me now take, in a few steps, several of the hurdles which I jumped over a period of 15 years. What are the possible interpretations of a function which maps sets into sets? Let f be a function defined on a certain class of sets which associates a set with each member of that class. The following interpretations are not meant to be exhaustive but suggestive of interpretations of f in the context: $p \epsilon fX$.

(1) X is close to p (topology, approximation)
(2) X implies p (logic f) (logic, inference)
(3) X generates p in the context f (algebra, geometry, analysis, formal languages)
(4) f is a computing machine and p is in the output if X is put in
(5) X protects (confines) p (e.g., fX is the interior of X, topology, fX is the convex hull of X)
(6) X is remote from p or X does not imply p (in the logic f) (geometry, graphs, kinship)
(7) X is a collection of freshmen, f is a university and $p \epsilon fX$ means p is a transformed or educated student from X
(8) p is a finished product from factory f where X comprises the raw materials

Thus it is apparent that there are many suggestive interpretations. Incidentally, all usual functions such as real-valued, complex-valued, and function-valued are embraced in the context of set-valued set-functions. The application here to universities and factories is an abstraction which is necessary for applying mathematics to any real system.

I thus conclude that I have no a priori reason for excluding from consideration any sort of function which may be selected by anyone. However I will restrict attention first to functions which map each subset of a "space" M with null set N into a set also contained in M. However, one has a choice of the set M, and the function f. Let the same selection apply to a "space" M_1 with null set N_1. Namely, choose any function g which associates with each set Y, contained in M_1, a subset of M_1.

Now consider any transformation t which maps each element of M into an element of M_1 and, by natural extension, t maps subsets of M into certain subsets of M_1. Then I shall define t to be (f,g)—continuous at p, $p \epsilon M$, provided $p \epsilon fX$ implies $tp \epsilon g(tX)$ which is the same form as proposed above. If f is a "logic" then t is continuous at p provided that from $X \underset{f}{\rightarrow} p$ there follows $tX \underset{g}{\rightarrow} tp$. That is, a continuous map preserves implication. It may also preserve remoteness, distance, protectivity, and so on.

Is this mild generalization irksome? Does it seem useless, a generalization for the sake of generalization or does it appear that continuity is about to become a more meaningful concept unifying and clarifying a wide range of hitherto diverse phenomena? The latter is the case. Continuity basically has nothing to do with geometric closeness except for purposes of specialization. You are "close" to relatives or friends who are thousands of miles away. You are close to finishing a job (in time); objects which may generate others in some context are close to them.

Conceded the theoretical possibility of enlarging the concept of continuity, of what use is it? The answers are increased applicability, transfer of information from one area to another, efficiency, and aesthetics. The applicability is now much greater and I may embrace such finitely described sets as computer programs, tapes of automata, finite groups, actual languages, as well as the infinite sets of algebras, analysis, geometry, and topology; computing machine systems and so on. Again, there have been many results achieved concerning forms of continuity. These often apply in this context to other specializations. I have thus established a communications channel among all the fields of mathematics and theories of systems and with comparatively little effort enriched each.

This brings us to efficiency. In how many thousands of places in the scientific and mathematical literature has it been proved that "a continuous function of a continuous function is continuous"? Why not prove it once and understand it more clearly? Why not use the general result to understand the individualism of the particular one?

Theorem 1. Let M, M_1, M_2 be three spaces. Let f, g, h be set-valued set-functions associated with M, M_1, M_2 respectively and let t_1 map M into M_1, let t_2 map M_1 into M_2. Then if t_1 is (f,g) continuous at $p \in M$ and t_2 is (g,h) continuous at $q = t_1 p \in M_1$ the composition $t = t_2 t_1$ mapping M into M_2 is (f,h) continuous.

Proof. Suppose $X \subseteq M$ and $p \in fX$. Then since t_i is (f,g) continuous at p, $q = t_1 p \in g(t_1 X)$. But since t_2 is (g,h) continuous at q, $t_2 q = t_2 t_1 p \in h t_2(t_1 X)$. Hence $p \in fX$ implies $tp = t_2 t_1 p \in h(t_2 t_1 X) = htX$. Then the composition $t = t_2 t_1$ is (f,h) continuous at p. : :

This formal proof, simple as it is, is clarified by the implication interpretation. Thus if from $X \vec{_f} p$ there follows $t_1 X \vec{_g} t_1 p = q$ and from $t_1 X \vec{_g} q$ there follows $t_2(t_1 X) \vec{_h} t_2 q = t_2 t_1 p$ then from $X \vec{_f} p$ follows $t_2 t_1 X \vec{_h} t_2 t_1 p$. That is, an implication-preserving map of an implication-preserving map preserves implication!

Engineers may prefer to think of information-preserving transformation rather than implication or closeness preserving. Insofar as my model is appropriate, i.e., if it can embrace the necessary details, it makes no particular difference what is preserved. This particular theorem can be presented to children, with examples. Its generality makes it possible to embrace simple examples. Its efficiency resides in the clarification and trivialization of the myriads of instances of the same result presented as distinct and needlessly consuming mental energy and time.

Very well, but can any theory be established within the framework I have proposed? Soon I will show in mathematical terms that the seeming freedom in the choices of f and g may lead to very stringent requirements on the continuous mappings compared to those achievable by restricting the character of these functions. Before proceeding to that task, we might ask: What is continuity? One of my goals was to embrace algebraic homomorphisms and topological homomorphisms in one framework. The construction given here fails to do the job. Continuity must be generalized in other directions to achieve the goal. Continuity, at this time, is perhaps best described as dual to invariance. Namely, a mapping is continuous with respect to whatever properties or relations it preserves or leaves invariant. Structure-preserving, information-preserving, closeness-preserving, angle-preserving, measure-preserving, distance-preserving, or form-preserving maps are continuous.

Coupled closely with this rigid interpretation of continuity is a more general approximate form so important in modeling. A model of an actual collection of systems is good if the mappings (into the model and from the model) preserve the information or structures considered important. For example, a translation of an article from French into English is an approximation to the original resulting from the translating transformation. If you say the translation is good, you have said that the translating was continuous because it preserved the important features of the original. Important stuff-preserving maps are continuous. Someone else may say the translation is not good. He considers something else to be important which is not preserved. One man's continuity is another man's discontinuity! Continuity is not intrinsic, it is relative to the demands of the moment.

In the case of translation, however, the composition theorem does not apply because of the approximate nature of the process. For example, Mark Twain has written an essay on the translation of his "Celebrated Jumping Frog of Calaveras County" into French and then back into English: it illustrates that not enough was preserved somewhere.

While the concept of continuity is simple, the ramifications are not and the anticipated symbolic maneuvers are discussed next.

II. FIRST LEVEL OF CONTINUITY

In this section I show that it is possible to construct a theory, important details of which will be missing for some time to come on the basis of my first generalization as presented. Let PM indicate the class of all subsets of a space M with null set N and let PM_1 indicate the class of all subsets of a space M_1 with null set N_2. I do not use $\emptyset$ as a universally empty set since it has no justification for serious mathematical work. Thus, the class PM of all subsets of M is dually the class of all supersets of N, the empty set. The class of all supersets of $\emptyset$, if I should grant such a monstrosity, would embrace PM, PM_1, and other things "too fierce to mention."

Now let F be the family of all functions f mapping PM into PM and let G be the family of all functions g mapping PM_1 into PM_1. I now discuss properties of F which apply also to G. In F there is induced an algebra corresponding to the set algebra in PM and an order relation corresponding to the inclusion relation in PM. Thus the union, $f_1 \cup f_2$, of two functions in F is defined by $(f_1 \cup f_2)X \equiv (f_1X) \cup (f_2X)$. The intersection $f_1 \cap f_2$ is defined by $(f_1 \cap f_2)X \equiv (f_1X) \cap (f_2X)$. For example, if f_1 and f_2 are called "logics" then, $p \in (f_1 \cup f_2)X$ if and only if $X\vec{f}_1\, p$ *or* $X\vec{f}_2 p$. Similarly, $p \in (f_1 \cap f_2)X$ if and only if $X\vec{f}_1 p$ and $X\vec{f}_2 p$. This is obviously related to switching circuits in engineering and computing. The inclusion $f_1 \subseteq f_2$ holds provided $f_1X \subseteq f_2X$ for all subsets X of M. This partial order relation among functions in F is analogous to the order relations in real-valued functions. I read $f_1 \subseteq f_2$ as "f_1 is contained in f_2," "f_1 is a subfunction of f_2," "f_2 contains f_1" or "f_2 is a superfunction of f_1". If $f_1 \subseteq f_2$ then from $X\vec{f}_1 p$ follows $X\vec{f}_2 p$. In terms of computing, f_1 has an output contained in the outputs of f_2 if the same input X is presented to both f_1 and f_2.

Now that switching and submachines have been introduced, we must also consider sequential machines or logics. The composition fg of an ordered pair (f,g) of functions in F is defined by $(fg)X \equiv f(gX)$. I treat composition as a form of multiplication indicated by juxtaposition. Obviously, this multiplication is not generally commutative but it is always associative, i.e., $f(gh) = (fg)h$ for every ordered triple (f,g,h) of functions from F. Note that I omitted the set X. I may do this by defining $f = g$ provided $f \subseteq g$ and $g \subseteq f$; i.e., $fX \equiv gX$ which is compatible with the partial ordering in F. In technical terms F is a semigroup with respect to composition because composition fg is defined for each ordered pair (f,g) of members of F, fg is in F, and composition is associative. In general, however, a function f in F has no inverse f^{-1} in F such that $f^{-1}f = ff^{-1} = e$ the identity function. Here $eX \equiv X$ and e is the identity with respect to composition; i.e., $ef = fe = f$.

Now composition, as suggested, corresponds to sequential machines. Thus $p \in fgX$ means that input X in machine g generates output gX which as an input to machine f yields p as an element of its output.

Similarly, $p \epsilon fgX$ is interpreted as $gX \ni p$; i.e., the set of all g implications of X implies p in the logic f. Parallel machines are not treated here since these correspond to functions of several variables, whereas in this case we are dealing with one set variable.

Now F has four distinguished functions—the identity e, the complement function c, the maximum function f_M and the minimum function f_N where $cX \equiv M \backslash X$, the complement of X in M, $f_M X \equiv M$ and $f_N X \equiv N$.

To maintain distinction between F and G, let e_1, c_1 respectively be identity and complement functions in G. Let there now be given a mapping t: $M \rightarrow M_1$. I first show the consequences of assuming t is (f,g) continuous at every $p \epsilon M$.

Theorem 2. Let t: $M \rightarrow M_1$ be (f,g) continuous where $(f,g) \epsilon F \times G$. Then if $f_1 = e \cup f$ and $g_1 = e_1 \cup g$, t is (f_1,g_1) continuous.

Proof. Suppose that t is (f,g) continuous. Then $p \epsilon fX$ implies $fp \epsilon g(tX)$ but $p \epsilon eX = X$ always implies $tp \epsilon tX = e_1(tX)$. Hence $p \epsilon X \cup fX = (e \cup f)X \equiv f_1 X$ implies $tp \epsilon tX \cup g(fX) = (e_1 \cup g)fX = g(tX)$. Accordingly, t is (f_1,g_1) continuous. : :

This theorem is stated before a more detailed discussion is undertaken. It could have been proved symbolically as follows. From $te = e_1 t$ and $tf \subseteq gt$ follows $te \cup tf \subseteq e_1 t \cup gt$ and by factoring t out, $t(e \cup f) = tf_1 \subseteq (e_1 \cup g)t = g_1 t$. Hence $tf_1 \subseteq g_1 t$. Note that t may be factored out on the left since it is additive, i.e., $t(X_1 \cup X_2) \equiv tX_1 \cup tX_2$.

Is this result of any use? Suppose $M = M_1$ is the set of real numbers and $fX \equiv gX$ is the set of limit points of X. Then a mapping t: $M \rightarrow M$ is (f,f) continuous if and only if it preserves limit points. Let $f_1 X \equiv X \cup fX$ as in the theorem. Then $f_1 X$ is the topological closure of X. The theorem says that all limit point-preserving maps are continuous in the usual sense, i.e., (f_1,f_2) continuous. The converse is not true since in topology all constant-valued transformations are continuous. However, such transformations do not preserve limit points. Those limit-point preserving maps, which should obviously be of interest in topology, are not included as a form of continuous mapping directly. Ordinary continuity allows destruction of information about limit points.

Now several definitions must be stated. Let t: $M \rightarrow M_1$ be a given transformation. Let $C(t) = \{(f,g): tf \subseteq gt\}$ be called the set of continuities of t. The problem now posed is: What is the character of $C(t)$? Does it contain many elements or few, does it have extreme elements; if it contains certain elements, which others are also present? The problem, intuitively, is this: Given any mapping t: $M \rightarrow M_1$ is it continuous in any way, does its set of continuities determine it, and how can we picture the set of continuities?

Theorem 3. Let $C(t)$ be the set of continuities of t. Then $(e,e_1)\epsilon C(t)$ so $C(t)$ is not empty. The following statements hold.

1. If $(f_i,g_i)\epsilon C(t)$ then $(\cup f_i, \cup g_i)$ and $(\cap f_i, \cap g_i)\epsilon C(t)$. That is, $C(t)$ is a complete vector lattice with respect to union and intersection.
2. If $(f,g)\epsilon C(t)$, then $(e\cup f, e_1\cup g)\epsilon C(t)$.
3. Let f_0 be the minimum inclusion-preserving function containing f and let g_0 be the minimum inclusion-preserving function containing g. Then $(f_0,g_0)\epsilon C(t)$ if $(f,g)\epsilon C(t)$.
4. If $(f,g)\epsilon C(t)$ and both f and g are inclusion-preserving, then $(f^2,g^2)\ldots(f^n,g^n)\ldots\epsilon C(t)$.
5. If (f_1,g_1) and $(f_2,g_2)\epsilon C(t)$ and g_1 is inclusion preserving, then $(f_1f_2, g_1g_2)\epsilon C(t)$.
6. Let $\bar{f}$, $\bar{g}$ respectively be the minimum closure functions containing f, g. Then $(\bar{f}, \bar{g})\epsilon C(t)$ if $(f,g)\epsilon C(t)$.
7. Let $f\epsilon F$ be any function. There exists a unique element (f_0, g_1) of $C(t)$ such that g_1 is the minimum function such that $(f,g_1)\epsilon C(t)$ and f_0 is the maximum function such that $(f_0,g_1)\epsilon C(t)$.
8. Let $g\epsilon G$. There exists a unique element (f_1,g_0) of $C(t)$ such that f_1 is the maximum function such that $(f_1,g)\epsilon C(t)$ and g_0 is the minimum function such that $(f_1,g_0)\epsilon C(t)$.
9. If s: $M\rightarrow M_1$ and $C(s)=C(t)$ then $s=t$; i.e., the continuities of a transformation determine it.

Proof. 1. Since $(f_i,g_i)\epsilon C(t)$, by definition $tf_i\subseteq g_it$ for each value of i. Hence $t(\cup f_i)=\cup(tf_i)\subseteq\cup(g_it)\equiv(g_i)t$ and t is $(\cup f_i, \cup g_i)$ continuous. Note that this proof uses the universal additivity of t. However, in general t is not intersective since two points may have the same image. But, t is inclusion preserving as applied to sets. Hence

$$t(\cap f_i)\subseteq\cap(tf_i)\subseteq\cap(g_it)=(\cap g_i)t \text{ and } (\cap f_i, \cap g_i)\epsilon C(t)$$

3. Part 2 is a restatement of Theorem 2. The existence of f_0 and g_0 is easy to establish since f_M is inclusion preserving and f_0 is the intersection of all inclusion-preserving maps containing f, for example. Also, $f_0X\equiv\cup\{fA\colon A\subseteq X\}$ is an algorithm for "computing" f_0 and $g_0Y\equiv\cup\{gB\colon B\subseteq Y\}$. Hence suppose $(f,g)\epsilon C(t)$. Then $tf_0X\equiv\cup\{tfA\colon A\subseteq X\}\subseteq\cup\{gtA\colon A\epsilon X\}=\cup\{gB\colon B\subseteq tX\}=g_0tX$. Hence $(f_0,g_0)\epsilon(Ct)$.

4,5. Let us prove that if (f_1,g_1) and $(f_2,g_2)\epsilon C(t)$ and g_1 preserves inclusion, then $(f_1f_2,g_1g_2)\epsilon C(t)$. I have $tf_1\subseteq g_1t$ and $tf_2\subseteq g_2t$. Hence $(tf_1)f_2\subseteq g_1tf_2$ on substituting f_2, on the right in $tf_1\subseteq g_1t$. But $tf_2\subseteq g_2t$. Hence, since g_1 is inclusion preserving $g_1(tf_2)\subseteq g_1(g_2t)$ and I have $tf_1f_2\subseteq g_1g_2t$ or, $(f_1f_2,g_1g_2)\epsilon C(t)$. Now, clearly, if f and g both preserve inclusion, I may apply the above conclusion with $f_1=f_2=f$ and $g_1=g_2=g$. Hence if $(f,g)\epsilon C(t)$ in these circumstances so is (f^2,g^2). Suppose $(f^n,g^n)\epsilon C(t)$. Then $tf^n\subseteq g^nt$. Substituting f on the right I find $tf^{n+1}\subseteq g^ntf$. But $tf\subseteq gt$ by assumption and since g^n is inclusion preserving, $g^n tf \subseteq g^n(gtf) \subseteq g^{n+1}tf$. Hence $(f^{n+1}, g^{n+1}) \in C(t)$. Note that the property that f preserves inclusion was not used. However, if g preserves inclusion, then in part 3 with f_0 as defined and $g_0 = g$ if $(f,g) \in C(t)$ then $(f_0,g) \in C(t)$.

6. Suppose $(f,g) \in C(t)$. Let F_0 be the minimum subfamily of F with these properties, $e \cup f_0$, where f_0 is as defined in part 3, is in F_0, and F_0 is closed with respect to composition and arbitrary union. Then the maximum function in F_0 is $\bar{f}$, the closure function of f. The closure $\bar{g}$ of g is similarly defined. However, with $e \cup f_0 = u$ and $e_1 \cup g_0 = v$, if $(f,g) \in C(t)$ then $(u,v) \in C(t)$ by parts 2 and 3. By part 4, $(u^n, v^n) \in C(t)$ since u and v preserve inclusion.

7. Let s: $M \to M_1$ and suppose $s \neq t$. I show $C(s) \neq C(t)$. Since $s \neq t$ there is a point $p_0 \in M$ such that $sp_0 = q_1 \neq tp_0 = q_2$. Let $g \in G$ be chosen as follows: $g(tM_1) = \{q_2\}$ and $gY = N_1$ for $Y \neq tM_1$. Let the function f_0 be chosen as in part 6, $f_0 = t^1 gt$ so $(f_0,g) \in C(t)$. Now if $tX \neq tM_1$ then $f_0X = N$. If $tX = tM_1$ then $f_0X = t^{-1}\{q_2\}$. Now $p_0 \in f_0 M_1$. But, $sp_0 = q_1 \notin g(tM_1) = \{q_2\}$ so $(f_0,g) \notin C(s)$. : :

Remarks

This theorem establishes many features of continuity. First, continuity is displayed as noninstrinsic. Each mapping has numbers of continuities. This fact is rather dimly realized in topology where deviation from usual sorts of continuity is possible but not encouraged for serious study. It is not possible, in the framework of topological continuity to characterize a transformation by the pairs of closures which make it continuous. This is because every constant function is continuous in topological spaces. However, the problem of what kinds of functions f, g one might consider and still imply uniqueness is not settled and has many ramifications.

It will be noticed that I leaned toward closure functions in Theorem 3. This was because they fitted my purposes best. However, one can consider, say, the maximum inclusion preserving subfunctions f_1 and g_1 of f and g and then iterate $e \cap f_1$ and $e_1 \cap g_1$ to show that if f^* and g^* are the maximum interior functions contained in f and g respectively and $(f,g) \in C(t)$ then (f_1,g_1), $(e \cap f, e_1 \cap g)$ and $(f^*,g^*) \in C(t)$. In topological terms, if f^* and g^* map sets into their interiors an (f^*,g^*) continuous mapping is an open mapping; i.e., preserves open sets. The strongest continuities of parts 7 and 8 determine all others in $C(t)$. If (f,g) is a strongest continuity of t then $u \subseteq f$, $v \supseteq g$ implies $(u,v) \in C(t)$.

Several other results may now be stated.

Theorem 4. Let t: $M \to M_1$ and suppose $(f,g) \in C(t)$. Then if g is a closure function $(\bar{f},g) \epsilon C(t)$ where $\bar{f}$ is the minimum closure function containing f.

This theorem, a corollary of Theorem 3, part 6, serves the useful purpose of saving work if g happens to be a closure and f does not. However, the asymmetry of the definitions of transformations does not attract us to the obverse and true statement that if f is a closure function and $(f,g) \epsilon C(t)$ then $(f,\bar{g}) \epsilon C(t)$ since $\bar{g} \supseteq g$ and hence generally $(f,\bar{g})$ continuity is weaker than (f,g) continuity.

Example

A continuity in the integers. Let M be the set of positive integers ≥ 2. Let $f \epsilon F$ be defined by $f\{n\} = \{n^2\}$ and if $m \neq n$ $f\{m,n\} = mn$. For all sets X which are neither singletons nor doublets fix $fX = N$. The problem now is to determine all those transformations t: $M \rightarrow M$ such that t is (f,f) continuous: I claim that t is (f,f) continuous if and only if t preserves multiplication. First consider $tf\{n\} \subseteq ft\{n\}$ now $tf\{n\} = tn^2$ and $f\{tn\} = (tn)^2$. Hence $tn^2 = (tn)^2$ and t necessarily preserves squares. Next if $m \neq n$ $tf\{m,n\} = t(mn) \subseteq ft\{m,n\} = tm \cdot tn$ (even if $tm = tn$). Hence $t(mn) = tmtn$ and t necessarily preserves multiplication. Now if t preserves multiplication, then $tfX \subseteq ftX$ since if X contains three or more numbers or is N, then $fX = N$ and $ftX \supseteq N$. But if X has one or two elements then $tfX = ftX$.

Let $p_1 \cdots p_k \cdots$ be the sequence of all prime numbers and let $n_1, \cdots, n_k, \cdots$ be any sequence of numbers in M. Let $n = p_1^{\alpha_1} \cdots p_k^{\alpha_k} \cdots$ be the decomposition of n and define $tn = n_1^{\alpha_1} \cdots n_k^{\alpha_k} \cdots$. In this fashion all multiplication-preserving maps of M into itself may be defined. For example, $n_k \equiv 2$ generates such a mapping and in this case $tn = 2^{\alpha_1 + \cdots + \alpha_k + \cdots}$. Note that all but a finite number of the α_k's are zero.

Now if I had used the closure $\bar{f}$ of f instead of f as defined, many other maps t would be continuous. For example, every constant-valued mapping t would be continuous. Thus the admission of functions more general than closures increases the descriptive power of continuity.

It is intuitively clear that in most cases where information or structure is to be preserved, constant-valued functions are not likely to be of service since they do not respond to the structure in the domain.

There are now numbers of ways of generating set-valued functions of some interest in various domains. I list a few. Let M be a linear vector space (e.g., the plane) and let poq be the open line segment between p and q where $p \neq q$. Then o is a binary operation in M which has certain sets as values. Let $f\{p,q\} = poq$ for $p \neq q$ and let $fX = N$ for all other sets. In this case the closure $\bar{f}$ of f is the convex hull function. If M is the line (real numbers) and t: $M \rightarrow M$ is (f,f) continuous, then t is either monotone increasing or monotone decreasing. However, since $\bar{f}\{p,q\} = [p,q]$ a closed interval, a function t is $(\bar{f},\bar{f})$ continuous provided it is monotonic and nondecreasing or nonincreasing. In particular, a constant function is $(\bar{f},\bar{f})$ continuous but is not (f,f) continuous. Continuities of this kind have not been investigated in higher dimensional spaces.

All algebraic operations lead to set-valued functions. However, product-preserving maps when the operations involved are not commutative cannot be characterized by the kinds of continuity discussed here so far. You may use composition of various set-functions to establish new kinds of continuity. For example, if fX is the convex hull of X, where M is the plane and uX is the topological closure of X, which mappings t, of M into itself, are (uf,uf) continuous? Which are (fu,fu) continuous? Again, suppose that M is the plane and fp is the open circular disk with center at p and unit radius and $fX = N$ for all set X which are not singletons. For what map t is $t: M \to M$ (f,f) continuous? For which t is t: $M \to M_1$ where M_1 is the real line with a similarly defined function $g\{a\} = (a-1,a+1)$, $gY = N$ for all Y not a singleton (f,g) continuous? This kind of continuity with the radius replaced by $\varepsilon > 0$, if you like, is important in approximation theory.

II. SECOND LEVEL OF CONTINUITY

Now let us look at the problem posed by the failure to include algebraic homomorphisms in the first generalization. In algebra a homomorphism is an operation-preserving mapping. I have indicated but not proved that commutative binary operations may be treated on the first level of continuity. What is the difficulty with preserving noncommutative operations? The answer is that there is no difficulty from the algebraist's viewpoint, the difficulty arises when I try to see algebraic homomorphism and topological homomorphism (continuous maps) as the same type of restriction.

Here, I discovered another way of representing continuity which embraces algebraic homomorphisms. This involves considering relations among elements and sets which involve as many components as needed. I will mention that relations involving an infinity of components are useful but, for simplicity, I assume here only a finite number. As before, let PM be the class of all subsets of a space M. Then an n-ary relation in PM is a set R of ordered n-tuples $(X_1,\cdots,X_n)$ of subsets of M. If R contains only n-tuples of one-element sets, it may be called an n-ary relation in M. Let me first mention an example. If f is a topological closure function in F then let $R = \{(X,Y): fX \cap Y \neq N\}$. That is $(X,Y) \in R$ if Y contains some $p \in fX$. This relation is equivalent to the closure function f since $fX \equiv \{p: (X, \{p\}) \epsilon R\}$. Preserving a relation R defined in this fashion is equivalent to topological continuity. That is, if $S = \{(Y_1,Y_2): gY_1 \cap Y_2 \neq N_1\}$ where g is a topological closure in a set M_1 and t: $M \rightarrow M_1$, then t is continuous provided $(X_1,X_2) \epsilon R$ implies $(tX_1,tX_2) \epsilon S$. Thus the suggestion now is that continuous maps preserve relations.

I adopt that suggestion. Now you may choose an n-ary relation R in PM and an n-ary relation S in PM_1 and I should say t: $M \rightarrow M_1$ is (R,S)-continuous provided $tR \subseteq S$; i.e., $(X_1,\cdots,X_n) \epsilon R$ implies $(tX_1,\cdots,tX_n) \epsilon S$. Of course, the composition of continuous mappings will again be continuous in a fashion analogous to that proved in Theorem 1. Does this device actually work for the arbitrary set-valued function continuity introduced before; i.e., does it generalize or is it different? Let f,g be arbitrary set-functions in F and G and define $R = \{(X,\{p\}): p \epsilon fX\}$. Let $S = \{(Y,\{q\}): q \epsilon gY\}$. Theorem 4, the mapping t: $M \rightarrow M_1$ is (f,g) continuous at p if and only if $(X,\{p\}) \epsilon R$ implies $(tX,\{tp\}) \epsilon S$. Hence t is (f,g) continuous if and only if t is (R,S) continuous.

The proof is obvious in view of the definitions. Thus relation-preserving maps embrace the first kind discussed. Suppose now M has a binary operation $\cdot$ and M_1 has a binary operation $\bigcirc$. Now let $R_0 = \{(p_1,p_2,p_1 \cdot p_2): p_1 p_2 \epsilon M\}$. $S_0 = \{(q_1,q_2,q_1 \bigcirc q_2): q_1,q_2 \epsilon M_1\}$.

Theorem 5. A mapping t from M to M_1 preserves multiplication if and only if t is (R_0,S_0) continuous.

Proof. Suppose t is (R_0,S_0) continuous. Then $(p_1,p_2, p_1 \cdot p_2) \epsilon R_0$ for each $p_1,p_2 \epsilon M$. Since t is (R_0,S_0) continuous then $(tp_1,tp_2,t(p_1 \cdot p_2)) \epsilon S_0$. Hence by definition of S_0, $t(p_1 \cdot p_2) = tp_1 \bigcirc tp_2$, which is to say t preserves multiplication or is a homomorphism. Conversely, if $t(p_1 \cdot p_2) = tp_1 \bigcirc tp_2$ for all $p_1,p_2 \epsilon M$, then $(p_1,p_2,p_1 \cdot p_2) \epsilon R_0$ implies $(tp_1,tp_2,t(p_1 \cdot p_2) \epsilon S_0$ and t is (R_0,S_0) continuous. : :

Note now that R_0 is the graph of the binary operation $\cdot$ and hence requires a ternary relation to represent it unless $\cdot$ is commutative. If $\cdot$ is commutative then $\{(\{p_1,p_2\},p_1 \cdot p_2)\}$ would be a binary relation which is equivalent to R_0. However, if $\cdot$ in M_1 were not a commutative operation, it is still possible that a mapping t of M into M_1 preserves multiplication. For this reason the ternary relation is preferable in representing all cases.

The comparison now with standard continuity is here, but to see it more clearly I can let $R = \{(X_1,X_2,X_3)$: for some $p_1,p_2 \epsilon M$, $p_1 \epsilon X_1$, $p_2 \epsilon X_2$ and $p_1 \cdot p_2 \epsilon X_3\}$ and similarly let S be a ternary set relation similarly generated by S_0 in PM. Still the result of Theorem 5 holds when R,S replaces R_0,S_0.

IV. CONCLUSION

Relation-preserving maps have very little theory behind them, although the number of special cases already considered is large. Elsewhere I have given a necessary and sufficient condition for connectedness of a general sort to be preserved in terms of binary relations among sets. The study of relation-preserving maps will provide more tools for systems theory and its applications. It will be found difficult to embrace many forms of structure-preserving maps in the contexts here presented. There are also the approximate and stochastic forms of continuity and certain clues to implicit forms of continuity in which mappings satisfy conditions which seem to resist being classified as continuous. In this latter category are Lipschitzian and differentiable real-valued functions.

5

FILTERS IN GENERAL

PRESTON C. HAMMER

I. INTRODUCTION

A filter is here defined as any device which accepts or passes certain elements in a set and rejects others. Thus a filter produces an ordered dichotomy in a set, the set of elements passed and its complement — those not passed. This definition is compatible with the common language concept and is a sweeping generalization of the topological definition.

It might be thought, at first sight, that it is a pleasant but useless pastime to give a definition of filters in such an obvious fashion. On the contrary, filters, so defined, embrace equations, inequalities, restrictions, and set properties, and they produce in collections all equivalence relations. Relations in general may be interpreted as filters but the interpretation of binary relations as filter collections is particularly useful, as I shall show in generalizing the filters of topology.

For example, let $y'=f(x,y)$ be an ordinary differential equation. Then each appropriate function which satisfies this equation is accepted by that filter. Let (x_0,y_0) be an initial point. Then the set of all functions u which satisfy $u(x_0)=y_0$ is accepted by that filter and those which pass both are accepted by the conjunction of both filters, which is then a filter. If we wish, we can consider the set of functions as being decomposed into possibly four equivalence classes by the two filters, viz., (accepted, accepted), (accepted, rejected) (rejected, accepted), (rejected, rejected). This gives a quaternary decomposition of the set of functions.

Suppose another requirement is made, for example, that the solution be analytic. This again is a filter and the conjunction of the three determines a filter and also an equivalence relation.

Thus the concept of filter enables us to embrace all limiting conditions in one framework. Two things I should remark. First, the generation of filters for particular objectives has a long mathematical, scientific, engineering, and cultural history. I do not claim to supplant or deprecate that work. On the contrary, I intend to unify the theory of filters and, incidentally, to produce a few new ones myself. Second, the theory here after a few preliminaries will settle down to generalization of topological filters, which generalizations are themselves specializations of the general concept.

II. GENERAL RESULTS CONCERNING FILTERS

Let M be a set with null set N. A filter in M is any device which accepts or rejects each element in M. Let A be the set of elements accepted by a filter. Then $cA=M\backslash A$ is the set of elements rejected by the filter. If no element is accepted, $A=N$ and if all elements are accepted $A=M$. Thus the ordered dichotomy (A,cA) of M indicates the result of filter action. If (B,cB) is an ordered dichotomy of M produced by another filter, then the conjunction filter produces an ordered dichotomy $(A\cap B,\ cA\cup cB)$ where $A\cap B$ is the set of elements accepted by both filters. The equivalence relation induced by the two (A,cA), (B,cB) dichotomous relations has equivalence classes $A\cap B$, $A\cap cB$, $cA\cap B$, and $cA\cap cB$.

One of the most obvious filter devices is a characteristic function which assigns to each element in M a 1 (accepted) or a 0 (rejected). Then the set $A\cap B$, for example, would correspond to 11 and $A\cap cB$ to 10.

Lemma 1. Every equivalence relation in a set M is generated by a collection of ordered dichotomies. If n is the minimum cardinal number such that $m\leq 2^n$ where m is the cardinal number of the set of distinct equivalence classes in M, then n is the minimum number of ordered dichotomies which refine to give the equivalence relation.

Proof. Let $\{A_i\}$ be the collection of equivalence classes for a given equivalence relation E. Then $\{(A_i,cA_i)\}$ is a collection of ordered dichotomies which induce E. This collection of dichotomies has the maximum cardinal number if we exclude using (M,N) or (N,M) if not necessary.

To see how to obtain the minimum number n, let C be a set of n distinct elements and let F be the family of all characteristic functions defined on C. Then F has 2^n functions. Since $2^n \geq m$, let there correspond to each A_i exactly one of the functions in F, say f_i. Now for $a \epsilon C$, let $B_a = \cup \{A_i : f_i(a) = 1\}$. Then $cB_a = \cup \{A_i : f_i(a) = 0\}$ and $\{(B_a, cB_a) : a \epsilon C\}$ is a collection of ordered dichotomies of M. Note that in effect each A_i is assigned a subset of C as its "coordinate" since f_i selects the subset $C_i = \{a : f_i(a) = 1\}$. That no smaller number than n will suffice is readily seen, since if $\{(D_j, cD_j)\}$ is a collection of ordered dichotomies generating E and the domain J of indices is of cardinal $n_0 < n$, then it can produce at most $2^{n_0} < m$ distinct sets by forming all intersections of the D_j and the cD_j. : :

It is now clear that there are exactly $|2^M|$ ordered dichotomies available in a set M (where $|X|$ designates the cardinal number of X) since this is the number of distinct subsets of M. Thus each filter in M must produce one of these. Of course ordered dichotomies are not necessary to give equivalence relations nor would it be necessary always to say a filter produces an ordered dichotomy since, in some cases only a dichotomy may be sought. However, the order does allow for more information such as providing binary representations of real numbers. The theory of filters in the degree of generality posed is here left for separate study. I turn now to special methods of describing filters.

III. BINARY RELATIONS AND FILTERS

Let R be a subset of $M_1 \otimes M_2$, where M_1 and M_2 are two sets with null sets N_1 and N_2 respectively. It is useful to think of R as a simple dictionary. Thus if M_1 is a collection of French words and M_2 a collection of English words, R might be a French-to-English dictionary where each French word is followed by a set of English words. The inverse relation $R^{-1} = \{(q,p) : (p,q) \epsilon R\}$, in this application would be the equivalent English-to-French dictionary. The subset R of $M_1 \otimes M_2$ may be called a relation in M_1 to M_2. Here I adopt the convention that $N_1 \times Y$ and $X \otimes N_2$ as well as $N_1 \otimes N_2$ are empty relations.

Now a binary relation, R, may be interpreted as a collection of filters if it is not empty. Thus if $u(p) = \{q : (p,q) \epsilon R\}$ then $u(p)$ is a subset of M_2 (which may be empty) and $u(p)$ may be interpreted as the subset of elements of M_2 accepted by (p,R). Dually, if $v(q) = \{p : (p,q) \epsilon R\}$, then $v(q)$ is the subset of elements of M_1 accepted by (R,q). There are, as I have mentioned, several ways of interpreting a collection of filters. I choose, in this case, to use the consequent function $u(p)$ and the antecedent function $v(q)$ of R to generate conjunction filters respectively in M_2 and M_1. Let $A \subseteq M_1$. Then I extend u by the definition $uA = \cap \{u(p) : p \in A\}$. Note that uA is a subset of M_2 and, following the usual convention, $\bar{u}A = M_2$ if $A = N_1$. Similarly, define v on subsets B of M_2 by $vB = \cap \{v(q) : q \in B\}$.

I will say uA is the set of elements (in M_2) accepted by the filter (A,R) and that vB is the set accepted by (R,B). Now note that the functions u and v as extended map PM_1 into PM_2 and PM_2 into PM_1 respectively where PX is the class of all subsets of X. Thus I may form alternate compositions uv which map PM_2 into itself and vu which maps PM_1 into itself.

Theorem 2. Let $R \subseteq M_1 \otimes M_2$ and let u, v be associated functions of R as described above. Then the following statements hold.

1. u and v are antitonic (inclusion-reversing) functions, i.e., if $A \subseteq A_1 \subseteq M_1$ then $uA \supseteq uA_1$.
2. $vuA \supseteq A$ and $uvB \supseteq B$ for each $A \subseteq M_1$ and $B \subseteq M_2$
3. $uvu = u$ and $vuv = v$ and hence $uvuv = uv$ and $vuvu = vu$ so uv and vu are closure functions.
4. For $A \subseteq M_1$ the set uA is the maximum subset of M_2 such that $A \otimes uA \subseteq R$.
5. For $B \subseteq M_2$ the set vB is the maximum subset of M_1 such that $vB \otimes B \subseteq R$.
6. Hence $vuA \otimes uA$ and $vB \otimes uvB$ are maximal product subrelations of R for each $A \subseteq M_1$ and each $B \subseteq M_2$.

Conversely, if $C \times D$ is a maximal product subrelation of R then $uC = D$, $vD = C$, $vuC = C$, and $uvD = D$.

Proof. 1. Since uA is the intersection of all sets $u(p)$ such that $p \in A$, if $A \subseteq A_1$ then $uA_1 \subseteq uA$ and u is antitonic. Similarly, v is antitonic.

2. Consider $v(uA) = \cap \{v(q) : q \in uA\}$. Now if $q \in uA$ then $(p,q) \in R$ for all $p \in A$, hence $p \in v(uA)$. If $uA = N_2$, then $vuA = M_1 \supseteq A$. Hence $vuA \supseteq A$ and similarly $uvB \supseteq B$ for $B \subseteq M_2$.

3, 4, 5, 6. By definition, uA is the maximum subset of M_2 such that $A \otimes uA \subseteq R$. Note that I may need to appeal to the convention that $A \otimes N_2$ is empty if $uA = N_2$. But then vB is, dually, the maximum subset of M_1 such that $vB \otimes B \subseteq R$. Let $B = uA$. Then vuA is the maximum subset of M_1 such that $vuA \otimes uA \subseteq R$. Since $A \otimes uA \subseteq R$ and A is maximal, it follows that $vuA \supseteq A$ and since vuA is maximal, $vuA \otimes uA$ is a maximal product subrelation of R. But then $vuA \otimes u(vuA) \subseteq R$ and $u(vuA)$ is the maximum subset of M_2 with this property. Hence $uvuA \equiv uA$, i.e., $uvu = u$. Dually, $vuv = v$.

Now uv is isotonic, enlarging ($uvB \supseteq B$ always), and idempotent.

Hence uv is a closure function mapping PM_2 into itself and, likewise, vu is a closure function mapping PM_1 into itself. : :

Definition. If $C \otimes D$ is a maximal product subrelation of R then (C,D) is called a dual filter pair (over R).

The above theorem guarantees existence (sometimes in trivial fashion) of dual filter pairs. Note that if (C, D) is a dual filter pair over R, then (C, R) accepts D and (R, D) accepts C.

It is most useful to interpret the above results for binary relations in the set of real numbers — where the binary relations may be pictured as subsets of the plane (see Figure 1).

Now I specialize to $M_1 = M_2 = M$ and $N_1 = N_2 = N$. Then if $R \subseteq M \otimes M$, R is often called a binary relation in M. All the preceding results and definitions hold. Now, however, it is possible to define another concept.

Definition. If A is a subset of M such that (A, A) is a dual filter pair over R, then A is called a self-dual filter in M (over R).

Lemma 3. A necessary and sufficient condition that a subset A of M be a self-dual filter over a binary relation R in M is that $A = uA = vA$. The dual filter pair (vB, uvB) over R generated by $B \subseteq M$ is self-dual if and only if $uvB = vB$. Hence B is a base for a self-dual filter if and only if $B \otimes B \subseteq R$ and $uvB \supseteq uB$.

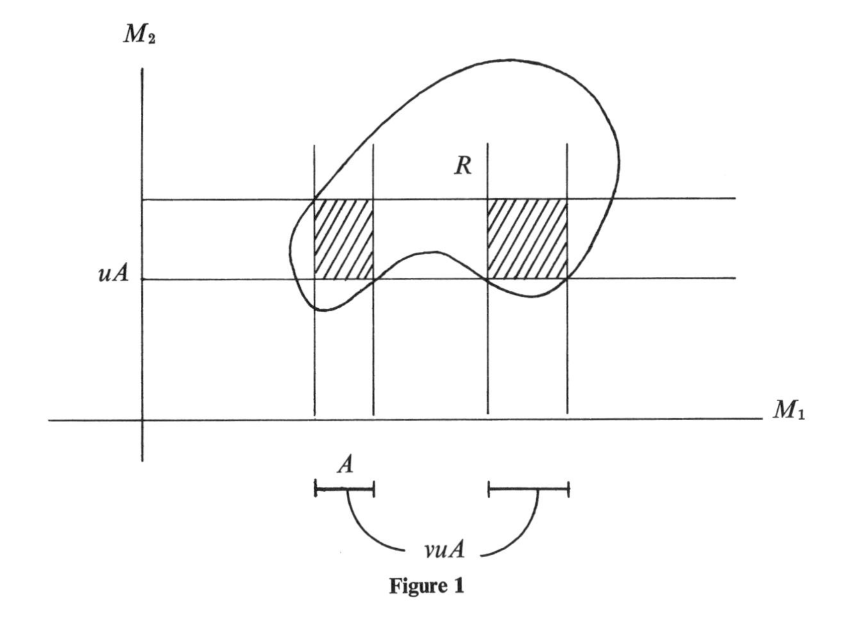

Figure 1

Proof. If $A=uA=vA$, then $vuA=A$ and hence $vuA\otimes uA=A\otimes A$ is a maximal product subrelation of R and A is a self-dual filter. Since $vB=uvB$ corresponds to the definition of self-dual filter, it is equivalent to it.

Suppose $B\otimes B\subseteq R$. Then $vB\supseteq B$ and since u is antitonic $uvB\subseteq uB$. But by assumption $uvB\supseteq uB$ and hence $uvB=uB$ so the conditions are sufficient. They are obviously necessary. : :

While self-dual filters may exist in nonsymmetric relations, it is obvious that since A being a self-dual filter implies $A\otimes A\subseteq R$, there will be most interest in restricting attention to symmetric relations R.

Definition. A relation R in M is said to be *relatively reflexive* provided $(p,q)\in R$ implies (q,q) and $(p,p)\in R$.

Before stating the next theorem, I observe that if R is a symmetric relation in M then $v=u$ and hence $uv=vu=u^2$ is a closure function and (u^2A, uA) and (uA, u^2A) are dual filter pairs over R. Hence if $u^2A=uA$, uA is self-dual filter in M.

Theorem 4. Let R be any nonempty relatively reflexive symmetric relation in M. Let $B_0\otimes B_0\subseteq R$ where $B_0\neq N$. There exists a self-dual filter A in M such that $A\supseteq B_0$.

Proof. Let $\{B_\alpha\}$ be a maximal ascending chain of subsets of M such that $B_1\supseteq B_0$ and $B_\alpha\otimes B_\alpha\subseteq R$. Let $A=\cup B_\alpha$. Then it is readily verified that $A\otimes A\subseteq R$ and hence A is the maximum member in $\{B_\alpha\}$. Hence I must show that $A\otimes A$ is a maximal product subrelation of R to establish that A is self-dual. Suppose $p\notin A$ and $(p\cup A)\otimes A\subseteq R$. Then, since R is symmetric $A\otimes(p\cup A)\subseteq R$ and since $A\supseteq B_0$ which is not empty, $(p,p)\in R$ since R is relatively reflexive. Hence $(p\cup A)\otimes(p\cup A)\subseteq R$ which contradicts the maximality of the chain $\{B_\alpha\}$. Hence A is a self-dual filter in R. : :

This theorem, based when necessary on the well-ordering axiom, shows the existence of self-dual filters for a variety of applications. Even so, these self-dual filters may not be easy to describe.

Corollary 5. Let R be a symmetric relatively reflexive relation in M and let (C, D) be a dual filter pair over R. Then if $C\supseteq D$, $D\neq N$, there exists a self-dual filter, A in M such that $C\supseteq A\supseteq D$. In other words, if one of a dual pair of filters over R contains the other, then there exists a self-dual filter which "separates" them.

Proof. If $C\supseteq D$ as described, then $D\otimes D\subseteq R$ and hence Theorem 4 applies with $B_0=D$ to yield a self-dual filter A such that $A\supseteq D$. However, since C is the maximum subset of M such that $C\otimes D\subseteq R$ and $A\supseteq D$, $A\otimes A\subseteq R$, it follows that $C\supseteq A$. Note also that if $C=A$ then $D=A$ and conversely. : :

IV. APPLICATIONS OF FILTERS TO CLASSES OF SETS

The foregoing apparatus is developed to prove the fundamental results associated with topological filters on the simpler level of relations in a set M. In topology a filter is a certain kind of collection of subsets of a space. The background relation R is submerged since usually only one is assumed.

Now the scene shifts to filters which are generated by binary relations in the power set PM of M and by various subclasses of PM. To be explicit I shall first define $R=\{(X,Y)\colon X\cap Y\neq N\}$, the collection of all pairs of subsets of M which are not disjoint. This is a symmetric relatively reflexive relation in PM. It misses being reflexive since $(N,N)\notin R$. Now the sets A and B of previous discussion are replaced by subclasses, say $\mathfrak{A}$ and $\mathfrak{B}$ of PM. The formal definition of $u\mathfrak{A}$, $v\mathfrak{B}$ is as before and all the properties proved hold.

Theorem 6. Let $R=\{(X,Y)\colon X\cap Y\neq N\}$. Then the pairs $(u^2\mathfrak{A}, u\mathfrak{A})$ and $(u^2\mathfrak{B}, u\mathfrak{B})$ are dual filter pairs over R. Moreover, in this context each of the classes $u\mathfrak{A}$, $u\mathfrak{B}$, $u^2\mathfrak{A}$, $u_2\mathfrak{B}$ are ancestral subclasses of PM. Moreover, there exists for each $\mathfrak{B}_0\subseteq PM$ such that $\mathfrak{B}_0\otimes\mathfrak{B}_0\subseteq R$ a self-dual filter $\mathfrak{A}$ such that $\mathfrak{A}\supseteq\mathfrak{B}_0$. Necessary and sufficient conditions for $\mathfrak{B}_0$ to be a base for a self-dual filter are that $\mathfrak{B}_0\otimes\mathfrak{B}_0\subseteq R$ and if A is a set such that $(A, Y)\epsilon R$ for all $Y\epsilon\mathfrak{B}_0$ then there exists a set $Y_0\epsilon\mathfrak{B}_0$ such that $A\supseteq Y_0$.

Proof. The first part is simply one application of previous results in view of the fact that R is symmetric and relatively reflexive. The ancestral property of the subclasses $u\mathfrak{A}$, $u\mathfrak{B}$, etc., arises since if $(X, Y)\epsilon R$ and $X_1\supseteq X$, $Y_1\supseteq Y$ then $(X_1, Y_1)\epsilon R$, i.e., R is "ancestral."

However, in this context the character of a base of a self-dual filter reflects the character of this application. That $\mathfrak{B}_0\otimes\mathfrak{B}_0\subseteq R$ is necessary follows from the definition. However, the property that $(A, Y)\epsilon R$ for all $Y\epsilon\mathfrak{B}_0$ implies that A contains a set in $\mathfrak{B}_0$ is distinctive. Suppose $\mathfrak{B}_0$ satisfies the two properties. Then $u\mathfrak{B}_0\supseteq\mathfrak{B}_0$ since $(X, Y)\epsilon R$ for all X, $Y\epsilon\mathfrak{B}_0$. Now suppose $(A, Y)\epsilon R$ for all $Y\epsilon\mathfrak{B}_0$. Then since $u^2\mathfrak{B}_0$ is an ancestral closure of $\mathfrak{B}_0$, and $A\epsilon\mathfrak{B}_0$, it is necessary, if $u^2\mathfrak{B}_0=u\mathfrak{B}$, that A be in the ancestral closure of $\mathfrak{B}_0$. This means that there is $Y_0\epsilon\mathfrak{B}_0$ such that $A\supseteq Y_0$. To prove sufficiency, since $u\mathfrak{B}_0=\{A\colon (A,Y)\epsilon R$ for all $Y\epsilon\mathfrak{B}_0\}$ if $A\epsilon u\mathfrak{B}_0$ implies there exists $Y_0\epsilon\mathfrak{B}_0$ such that $A\supseteq Y_0$ then $A\epsilon u^2\mathfrak{B}_0$ and $u^2\mathfrak{B}_0\supseteq u\mathfrak{B}_0$. But $u\mathfrak{B}_0\supseteq\mathfrak{B}_0$ and hence $u^2\mathfrak{B}_0\subseteq u\mathfrak{B}_0$ since u is antitonic. Therefore $u^2\mathfrak{B}_0=u\mathfrak{B}_0$ and $\mathfrak{B}_0$ is a base for a self-dual filter in PM. : :

V. TOPOLOGICAL FILTERS

In topological space theory the neighborhood filters are required to be ancestral classes. However, I prefer to use the usual bases as filters since they determine which sets are accepted. The binary relation $R = \{(X, Y): X \cap Y \neq N\}$ is submerged in topology. A filter base of neighborhoods of a point p in a topological space is a subclass $\mathcal{V}_0$ of PM with the following properties. (1) Each set in $\mathcal{V}_0$ contains p and $\mathcal{V}_0 \otimes \mathcal{V}_0 \subseteq R$, i.e., $X \cap Y \neq N$ if $X, Y \epsilon \mathcal{V}_0$. Moreover, if $X, Y \epsilon \mathcal{V}_0$ then there is $Z \epsilon \mathcal{V}_0$ such that $X \cap Y \supseteq Z$. The class $u\mathcal{V}_0 = \mathcal{C}(p)$, say, is the class of all sets X such that p is in the closure of X. I call $\mathcal{C}(p)$ the class of convergents of p. Then the dual filter pair is $[\mathcal{C}(p), \mathcal{V}(p)]$ where $\mathcal{V}(p)$ is the ancestral closure of $\mathcal{V}_0$. Now in topological spaces $\mathcal{V}(p) \otimes \mathcal{V}(p) \subseteq R$ and hence $\mathcal{C}(p) \supseteq \mathcal{V}(p)$ and usually the inclusion is a proper one.

Hence there exists a self-dual filter $\mathfrak{A}(p)$ such that $\mathcal{C}(p) \supseteq \mathfrak{A}(p) \supseteq \mathcal{V}(p)$. A self-dual filter is called an ultrafilter. A topological space then has for each $p \epsilon M$ a dual filter pair $[\mathcal{C}(p), \mathcal{V}(p)]$. If f and g are functions mapping PM into itself defined by $p \epsilon fX$ provided $X \epsilon \mathcal{C}(p)$ and $p \epsilon gY$ provided $Y \epsilon \mathcal{V}(p)$, then f is the Kuratowski closure function of the topology and $g = cfc$ is the dual interior function. In another paper I propose to investigate various convergence conditions in the more general isotonic spaces. I note in passing that the equivalence relation induced in PM by the neighborhood filter family is given by $A \sim B$ provided $fA = fB$. Of course the equivalence relation induced by the convergent filters is that determined by $gA = gB$.

VI. OTHER SET RELATIONS AND FILTERS

The aside on topological filters was introduced to show that the results here relate to those. With some important exceptions, topological filters are not suitable for application to constructive system theories because they generally do not apply to algebras and automata or finitary systems. Unfortunately, there are too many concepts now available to develop to any extent. I first merely mention that assigning to each $p \epsilon M$ a subclass $\mathcal{V}_0(p)$ of PM leads to significant generalizations of topological spaces where $R = \{(X, Y): X \cap Y \neq N\}$. Then there will be a collection of dual filter pairs $\{\mathcal{C}(p), \mathcal{V}(p)\}$ which are ancestral and $\mathcal{V}(p)$ is the ancestral closure of $\mathcal{V}_0(p)$. Again, associated functions f, g are defined and $cfc = g$. Moreover, f and g are isotonic but otherwise nothing more explicit can be said without further restrictions.

Of particular interest to me now are self-dual filters and more general relations. Let me suppose G is a set partially ordered by $\leq$ and that μ

is a function mapping each subset of M into an element of G. I shall assume μ is order-preserving, i.e., $X \subseteq Y$ implies $\mu X \leq \mu Y$. Then I shall define various relations R_m where $m \in G$ by $R_m = \{(X,Y)\colon \mu(X \cap Y) \geq m\}$. Such a relation R_m is symmetric, relatively reflexive, and ancestral. Hence the stage is set for the existence of self-dual filters and for ancestral classes of accepted sets.

Let me deviate a moment to give several instances of such functions μ since the abstraction may appear futile.

1. Let $\mu X = |X|$ the cardinal number of X. Then $R_1 = \{(X,Y)\colon |X \cap Y| \geq 1\}$ and this is the same relation used in topological filters. If I use $R_2 = \{(X,Y)\colon |X \cap Y| \geq 2\}$ then with a usual neighborhood base of $p \in M$, of a topology $\mathcal{C}_2(p)$ is the class of all sets X such that p is a limit point of X. This R_2 is well worth exploring.
2. Let M be a metric space and let $\mu X =$ diameter of $X = \sup\{d(p,q)\colon p,q \in X\}$. Then for $m > 0$, $R_m = \{(X,Y)\colon \text{diam}(X \cap Y) \geq m\}$. Clearly if a class $\mathfrak{B}_0$ of subsets of M is going to nontrivially provide a class $\mu\mathfrak{B}_0$ of subsets, then $Y \in \mathfrak{B}_0$ will require diam $Y \geq m$. This is a rather complex relation situation but it is important in dealing with certain special subsets of metric spaces.
3. Let M be a set and let μX be the exterior measure of X. Then $m > 0$ gives rise to nontrivial relations R_m important, in special cases, to convexity theory.
4. Let M be a product space. Let $\mu X = \{\mu_1\pi_1 X, \cdots, \mu_n\pi_n X\}$, where $\pi_i X$ is a projection of X to a coordinate plane, say, and $\mu_i\pi_i X$ is any suitable measure of $\pi_i X$ in the plane. This gives an example of vector-valued (partially ordered) measures.
5. Let M be a set covered by a countable class $\mathcal{C}$ of sets to each of which is assigned a positive real number so that their sum is finite. Let $\mu X = \sum_i \{r_i\colon Y_i \subseteq X,\ Y_i \in \mathcal{C}\}$. Alternately, let $\mu X = \inf\{\sum_i r_i\colon \cup Y_i \supseteq X,\ Y_i \in \mathcal{C}\}$. These measures μ may be related to packing and covering problems.
6. Let M be well ordered: $\{p_\alpha\}$. Let $\mu X = \sup\{\alpha\colon p_\alpha \in X\}$. Then $R_m = \{(X,Y)\colon \mu(X \cap Y) \geq m\}$. And $(X, Y) \in R_m$ if and only if there is a $p_\beta \in \{p_\alpha\}$ in common to X and Y and $\beta \geq m$.

These give a few of the examples which can be generated en masse with some thought. I turn first to the explicit case where $\mu X = |X|$ and m is a positive integer.

Theorem 7. Let m be a positive integer and let $R_m = \{(X,Y)\colon |X \cap Y| \geq m\}$. Let M be a set of $2k - m$ elements where $2k - m \geq 0$.

Then the class $\mathfrak{B}_0$ of all k-tuples from M is the base of a self-dual filter in M, and if M is a subset of a space E, then the ancestral closure of $\mathfrak{B}_0$ in PE is a self-dual filter in E over R_m.

Proof. If X, $Y \epsilon \mathfrak{B}_0$ then $|X \cap Y| \geq m$ since $|X| = k = |Y|$ and M has only $2k-m$ elements. Hence $\mathfrak{B}_0 \times \mathfrak{B}_0 \subseteq R_m$. Next, suppose $|A \cap Y| \geq m$ for every $Y \epsilon \mathfrak{B}_0$. I claim that $A \supseteq Y_0$ for some $Y_0 \epsilon \mathfrak{B}_0$. Suppose A does not contain any set in $\mathfrak{B}_0$. Then since $A \subseteq M$, $|A| < k$ since $\mathfrak{B}_0$ contains all k-sets from M. But then some set X_0, say, in $\mathfrak{B}_0$ contains A. Let X_1 contain all points, at least $2k-m-(k-1) = k-m+1$ in number, in cA and enough elements of A to give k elements. Hence $X_1 \epsilon \mathfrak{B}_0$, but then $|A \cap X_1| < m$, which was assumed not the case. Hence A contains some subset Y_0 of $\mathfrak{B}_0$ and the ancestral closure $\mu\mathfrak{B}_0$ of $\mathfrak{B}_0$ is a self-dual filter. : :

The theorem not only shows that self-dual filters exist but shows that bases for self-dual filters over R_m comprised of various classes of k-tuples exist for each positive integer m. Let us consider explicitly the case $m=1$. Then with $k=1, 2, 3$ sets of k-tuples in a base $\mathfrak{B}_0$ number 1, 3, and 10 respectively using the mechanism in the theorem. However, 10 is not the minimum number of triples which can provide a base for a self-dual filter over R_1. One example results from taking a seven-element set, say $M = \{1,2,3,4,5,6,7\}$ and taking a set of triples which form lines of a finite projective plane of order 2. For example, $\mathfrak{B}_0 = \{1,2,3\}\ \{1,4,5\}\ \{1,6,7\}\ \{2,4,6\}\ \{2,5,7\}\ \{3,4,7\}\ \{3,5,6\}$ is a base for a self-dual filter. I conjecture that seven is the minimum number of triples which can form a base for a self-dual filter over R_1.

At first this result raised the hope that the finite projective planes, when they exist, would provide bases for self-dual filters over R_1. For $k=4$, the number $2k-1$ is 7 and the set of all quadruples from a 7-element set is, according to the theorem, a base for self-dual filter over R_1. Now the number of elements in a finite projective plane of order 3 is 13 and there are 13 quadruples of points which comprise its lines. Such a set of quadruples is $\{1,2,3,4\}$, $\{1,5,6,7\}$, $\{1,8,9,10\}$, $\{1,11,12,13\}$, $\{2,5,8,11\}$, $\{2,6,9,12\}$, $\{2,7,10,13\}$, $\{3,5,10,12\}$, $\{3,7,9,11\}$, $\{4,5,9,13\}$, $\{4,7,8,12\}$, $\{4,6,10,11\}$. Now, however, this set is not a base of a self-dual filter over R_1. The reason is that there exist sextuples which intersect each but do not contain any one. For example $\{1,5,8,9,11,13\}$ intersects each of the thirteen but does not contain one. Moreover, although I did not establish a count, it appears that more than 35 sets will be required, including these 13, to form a base with at least four elements.

Problems

1. What is the minimum cardinal number of self-dual filter bases over R_1 provided the minimum cardinal sets have at least k-elements? over R_m?
2. Under what circumstances is a self-dual filter base comprised solely of k-tuples beyond the ones described in Theorem 8?
3. Every self-dual filter over R_1 divides the power set into two equal cardinal classes. Let M be a set with a fixed number n of elements. What is the range of cardinal numbers of the self-dual filters over R_m? I have shown that it is not constant; for $m=2$.
4. Starting from the class of lines of an $(k-1)^{\text{th}}$ order finite projective plane, what is the minimum number of additional sets it is necessary to adjoin to form a self-dual filter base for $n \geq 3$?

Remarks

I have shown the existence of a self-dual filter over R_m between $u\mathfrak{B}_0$ and $u^2\mathfrak{B}_0$ when $\mathfrak{B}_0 \otimes \mathfrak{B}_0 \subseteq R_m$ since $u\mathfrak{B}_0 \supseteq \mathfrak{B}_0$ and $u^2\mathfrak{B}_0 \subseteq u\mathfrak{B}_0$ in that case and $u^2\mathfrak{B}_0 \supseteq \mathfrak{B}_0$. Thus, starting from any class, $\mathfrak{B}_0$ such that X, $Y \epsilon \mathfrak{B}_0$ implies $|X \cap Y| \geq m$ leads to the existence of a class $\mathfrak{B} \supseteq \mathfrak{B}_0$ which is a base for a self-dual filter. However, in general the class $\mathfrak{B}$ will not be uniquely determined.

VII. NONANCESTRAL FILTERS

The designation of filtering action is not specified by simply giving a relation since a relation may be used in numerous ways to provide ordered dichotomies: for example, in topology the relation $R = \{(X,Y)\colon X \cap Y \neq N\}$ is used but it is usually not mentioned as a relation. The properties of neighborhood filter bases and of neighborhood filters are deemed the essential features. Thus the domain of the classes of sets may be restricted in various ways. Again, for a given class of sets $\mathfrak{B}_0$ we might use $\{X\colon (X,Y) \notin R \text{ for some } Y \epsilon \mathfrak{B}_0\}$ as the class of accepted sets.

So far, the special relations among sets considered here are ancestral. But the general theory, introduced earlier, makes no such restriction. Hence I may well wish to consider, for example, a binary relation in PM of the form $S_1 = \{(X,Y)\colon |X \cap Y| = 1\}$. Then for a given subclass $\mathfrak{B}_0$ of PM, $u\mathfrak{B}_0$ is the class of all subsets X of M which intersect each $Y \epsilon \mathfrak{B}_0$ in exactly one point. For one example consider $\mathfrak{B}_0$ to be the collection of equivalence classes A_i in M determined by an equivalence relation in M. Then $u\mathfrak{B}_0$ here would be the class of all sets which contain one element from each A_i, i.e., $u\mathfrak{B}_0$ is the class of representative sets of the equivalence relation. Then, $u^2\mathfrak{B}_0$ is $\mathfrak{B}_0$ and I have a nice dual filter pair $(u\mathfrak{B}_0, \mathfrak{B}_o)$.

To characterize finite projective planes in a filtering context, I need irreflexive relations. Let

$$T_1 = \{(X,Y) \colon X \neq Y,\ |X \cap Y| = 1\}$$

be a binary relation in PM. Let a modified Cartesian product, $*$, be defined by

$$A * B = \{(p,q) \colon p \neq q,\ p \epsilon A,\ q \epsilon B\}.$$

Lemma 8. Let $T_m = \{(X,Y) \colon X \neq Y,\ |X \cap Y| = m\}$ be a binary relation in PM. Let $\mathfrak{B}_0$ be a subclass of PM such that $\mathfrak{B}_0 * \mathfrak{B}_0 \supseteq T_m$. Then there exists a maximal superclass $\mathfrak{A}$ of $\mathfrak{B}_0$ such that $\mathfrak{A} * \mathfrak{A} \subseteq T_m$ and if $\mathcal{C} \supseteq \mathfrak{A}$ $\mathcal{C} \neq \mathfrak{A}$ then $\mathcal{C} * \mathfrak{A} \not\subseteq T_m$, $\mathfrak{A} * \not\subseteq T_m$.

Proof. Let $\{\mathfrak{B}_\alpha\}$ $\alpha \geq 0$ be a maximal ascending chain of subclasses of PM such that $\mathfrak{B}_\alpha * \mathfrak{B}_\alpha \subseteq T_m$. Let $\mathfrak{A} = \cup \mathfrak{B}_\alpha$. Now if $X, Y \epsilon \mathfrak{A}$, $X \neq Y$ then for some value, β of α X, $Y \epsilon \mathfrak{B}_\beta$ since $\{\mathfrak{B}_\alpha\}$ is an ascending chain. Hence $(X,Y) \epsilon T_m$ and $\mathfrak{A} * \mathfrak{A} \subseteq T_m$.

To show that $\mathfrak{A}$ is a maximal $*$-subrelation of T_m, suppose to the contrary that $Z \notin \mathfrak{A}$ and $(\mathfrak{A} \cup \{Z\}) * \mathfrak{A} \subseteq T_m$. Since T_m is symmetric, then $\mathfrak{A} * (\mathfrak{A} \cup \{Z\}) \subseteq T_m$. But then $(\mathfrak{A} \cup \{Z\}) * (\mathfrak{A} \cup \{Z\}) \subseteq T_m$ since $(Z,Z) \notin T_m$. This contradicts maximality of $\{\mathfrak{B}_\alpha\}$. Hence $\mathfrak{A} * \mathfrak{A}$ is a maximal $*$-subrelation of T_m. : :

Theorem 9. Let M be comprised of $n^2 - n + 1$ elements where $n \geq 2$. Let $T_1 = \{(X,Y) \colon X \neq Y,\ |X \cap Y| = 1\}$. Then a subclass $\mathfrak{B}$ of PM is comprised of the lines of a finite projective plane of order $n-1$ provided $\mathfrak{B} * \mathfrak{B}$ is a maximal $*$-subrelation of T_1, each set in $\mathfrak{B}$ has n elements, and the number of sets in $\mathfrak{B}$ is $n^2 - n + 1$.

Proof. From the definition of finite projective planes, it is necessary that each pair of distinct lines intersect in one point and the number of lines must equal the number of points—$n^2 - n + 1$ for a plane of order $n-1$. Moreover, the number of points on each line is n. What is left to show is that $\mathfrak{B} * \mathfrak{B}$ is a maximal $*$-subrelation of T_1 when $\mathfrak{B}$ is the class of lines of a finite projective plane of order $n-1$.

Suppose $A \subseteq M$, $A \notin \mathfrak{B}$ and $|A \cap Y| = 1$ for each $Y \epsilon \mathfrak{B}$. Then A must have at least n elements. If A has exactly n elements, then $A \epsilon \mathfrak{B}$ since every n-element subset of M which meets every line in $\mathfrak{B}$ is one of the lines. Hence A must have at least $m+1$ elements. But then A must have at least two elements in common with one of its lines, say Y_0. Then $|A \cap Y| \geq 2$ and $(A, Y_0) \notin T_1$. Hence $\mathfrak{B} * \mathfrak{B}$ is a maximal $*$-subrelation of T_1. : :

Remarks

Lemma 8 gives the existence of various "self-dual" filters in *PM* over T_m. Theorem 9 merely embeds projective planes, when they exist, in the context of such self-dual filters. It is known that projective planes of order 6 ($n=7$) cannot exist. Projective planes of order 10 have not been shown to exist. Then Lemma 8, applied to T_1 with $\mathfrak{B}_0$ comprised of certain sets of 11 elements each, says that there exists a maximal superclass $\mathfrak{A}$ of $\mathfrak{B}_0$ such that $\mathfrak{A}*\mathfrak{A}\subseteq T_1$ (when $\mathfrak{B}_0*\mathfrak{B}_0\subseteq T_1$). If the finite projective plane of order 10 does not exist, these classes $\mathfrak{A}$ can be considered approximations to a finite projective plane of order 10.

If we add to the relation T_1, the added requirement $|X|=|Y|=11$, we would have in the classes $\mathfrak{A}$ the "best" approximations to finite projective planes of order 10.

Problem

Let $\mathfrak{B}$ be the following class of eleven quintuples: {1,2,3,4,5} {1,2,6,7,8}, {1,3,5,9,10}, {1,4,7,9,11}, {1,5,8,10,11}, {2,3,8,9,11}, {2,4,6,10,11}, {2,5,7,9,10}, {3,4,7,8,10}, {3,5,6,7,11}, {4,5,6,8,9}. Prove or disprove that $\mathfrak{B}*\mathfrak{B}$ is a maximal $*$-subrelation of T_2. Note that each pair of sets from $\mathfrak{B}$ intersect in two elements.

VIII. CONCLUDING REMARKS

I believe I have gone far enough to accomplish the objectives of this paper. The terminology of filters provides a unification of broad areas of mathematics. Although filters may be given an even broader role than I have discussed, I think of them as related to the more or less quiescent role of selection or inspection. Then a function may be indicated as a collection of filters, being representable by its graph, a binary relation. However, I do not think of filters in the conceptual structure of transformation, or of generation which may be represented also by functions. Filters are go–no go devices, inspection criteria, or selection devices. Nevertheless, filter conceptualization is more appropriate for abstract functions than is the term "transformation."

Although I have not pursued the various applications in this paper, I have elsewhere shown how extension of the topological neighborhood filter concept provides a fundamental and powerful tool in the study of convexity (Hammer 1963, 1965, 1967). The reason I seized on the filter terminology was that situations arose in which the neighborhood terminology was not suitable. I found myself dealing with classes of sets which were not associated with points, for example. While in the presence of a suitable relation these were reasonably called filters, they could not reasonably be called neighborhoods of anything.

ACKNOWLEDGMENT

This work was partially supported by the National Science Foundation GP7077 funds and by the Ordnance Research Laboratory, U.S. Navy, the Pennsylvania State University.

REFERENCES

Bourbaki, N. 1940. "Topologie Generale," *Actualities Sci. Ind.* 858 (Filters).

Gastl, G.C. and P.C. Hammer. 1967. "Extended Topology: Neighborhoods and Convergents," *Proc. Coll. on Convexity,* Copenhagen, 1965, pp. 104–116.

Hammer, P.C. 1963. "Semispaces and the Topology of Convexity," *Proc. of Symposium on Pure Math.* Providence, Vol. 7.

Hammer, P.C. 1965. "Extended Topology: Caratheodory's Theorem on Convex Hulls," *Rend. Circ. Mat. Palermo* Ser. 2 **14**: 34–42.

Hammer, P.C. 1967. "Isotonic Spaces in Convexity," *Proc. on Coll. Convexity,* Copenhagen, 1965, pp. 133–141.

Kelly, J. L. 1955. *General Topology.* Van Nostrand, New York, p. 83.

Mann, H. B. 1949. *Analysis and Design of Experiments.* Dover Publications, New York.

Ryser, H. J. 1963. *Combinatorial Mathematics.* Buffalo Math. Assoc. of America. Distributed by John Wiley & Sons, New York: Carus Monograph 14.

6

APPROXIMATION SPACES

PRESTON C. HAMMER

I. INTRODUCTION

In this essay I propose to give a definition of mathematical approximations which makes general sense. Approximation theory has been enmeshed in the details of normed linear spaces, and has been allowed to become so by numerical analysts who should long since have detected the inadequacy of such systems to embrace their own work.

As usual, the real world approximations are too difficult to embrace completely in a mathematical description. Thus a system of differential equations may be used to represent the motions of objects in the solar systems along with certain measurements. While it is possible in principle to discuss how well a numerical or other approximation to the solution of these equations approximates the solution, the discussion of how well the mathematical system actually represents the solar system is much more difficult. For one thing, we steer away from introducing complex structures where simple ones seem to do as well. For another, the objectives must be kept in mind in order to get some standard. Is the moon, for example, to be treated as a separate object from the earth? How close should predictions and measurements be?

Another example arises in elemental translation, say from French to English. If I have a simple French-to-English dictionary in which each French word is followed by a collection of English words, then I have an approximation system. Suppose I were given a set of French words and the task of replacing each by one English word. Now comes the question of criteria of goodness of fit. The selection of best words is not easily described and this is the crux of this approximation problem. Actual translation of text, of course, is rendered more difficult since sentences and larger subsets must be translated rather than given as merely words. Sentence structure is important and so are connotations.

All models and simulations of systems are approximations. The confusion of the merit of an approximation with its existence as an approximation is standard and regrettable. I first realized this in considering numerical solutions of the differential equations of type

$$y' = f(x,y), \text{ initial point } (x_0, y_0)$$

Any appropriate function may be thought of as an approximation and it is the case that a criterion can be given in which it is the best, if only by declaring it so! There is no a priori measure of goodness of fit which is satisfactory for all purposes.

An approximation system then will become an approximation space when I place a preference relation as the set of approximants. This preference relation will simply be an order relation (i.e., a transitive binary relation) in the set of approximants. I may simply, for example, divide the approximants into two classes—good and bad—and this is an order relation. I have benefited by discussing approximations with Professor D. G. Moursund. Professor S. G. Mrowka suggested that I write the system as a ternary relation directly. Professor Moursund originally favored calling what I here call a uniform approximation space an approximation space.

II. APPROXIMATION SPACES

Relative to a given object p let $v(p)$ be a set of objects allowable as approximants of p. Normally, I will assume $v(p)$ is not empty although in contexts it may be allowable to have $v(p)$ empty—for example, a French word which has no English approximant. Now, I have already committed myself in choosing the set $v(p)$ since I will think of that set itself as being one best approximant of p. In particular, of course, we do select a set of allowable approximants of p. Naturally, saying $v(p)$ is a best set does not say that it is good.

Now in $v(p)$, I assume an order relation $T(p)$, i.e., $T(p)$ is a transitive binary relation. I say "q_1 is at least as good an approximant of p as q_2" provided $(q_1,q_2) \in T(p)$. Thus $T(p)$ establishes comparative desirability when it is present. In this context I may as well also assume $(q,q) \in T(p)$ for all $q \in v(p)$. Hence $T(p)$ is a reflexive order relation. A binary relation

$T(p)$ in $v(p)$ is called total provided $q_1,q_2 \in v(p)$ implies $(q_1,q_2) \in T(p)$ or $(q_2,q_1) \in T(p)$.

Note that if $q_1,q_2 \in v(p)$ and (q_1,q_2), $(q_2,q_1) \in T(p)$ then q_1 and q_2 are equally good approximations to p. Since this situation often happens, I do not require $T(p)$ to be antisymmetric.

Now, $\mathfrak{A}(p) = \{(p,q_1,q_2): q_1,q_2 \in v(p): (q_1,q_2) \in T(p)\}$ gives an approximation space for the one object p. A set $A \subseteq v(p)$ is called a best approximant of p if and only if to each $q \in v(p)$ there is a $q_0 \in A$ such that $(q_0,q) \in T(p)$. Now let E be a set of objects p to be approximated. Let $v(p)$ and $T(p)$ be given for each $p \in E$. Here $T(p)$ is a reflexive order relation in $v(p)$. Then

$$\mathfrak{A} = \{(p,q_1,q_2): p \in E, (q_1,q_2) \in T(p)\}$$

is called an *approximation space*. Letting $V = \cup \{v(p): p \in E\}$, then the approximation space $\mathfrak{A}$ is a subset of $E \otimes V \otimes V$, i.e., a ternary relation. A set $A \subseteq V$ is a *best approximant* of p provided $A \cap v(p)$ is a best approximant of p. Two approximation spaces in (E,V), $\mathfrak{A}_1$ and $\mathfrak{A}_2$ are *equivalent* provided that for each $p \in E$ the class of best approximants $A \subseteq V$, is the same for both $\mathfrak{A}_1$ and $\mathfrak{A}_2$.

Let $\mathfrak{A}$ be a given approximation space determined by $v(p)$, $T(p)$, $p \in E$. Let $E_0 \subseteq E$ and $V_0 \subseteq V$. Let $v_0(p) = v(p) \cap V_0$ for $p \in E_0$ and let $T_0(p) = \{(q_1,q_2): q_1,q_2 \in v_0(p), (q_1,q_2) \in T(p)\}$. Then the approximation subspace $\mathfrak{A}_0$ of $\mathfrak{A}_1$ determined by E_0, V_0 is $\mathfrak{A}_0 = \{(p,q_1,q_2): p \in E_0, (q_1,q_2) \in T_0(p)\}$. In this case, note that if $V_0 = \cup \{v(p): p \in E_0\}$, I would have the same criterion of approximation as in $\mathfrak{A}$ for each $p \in E_0$ $T_0(p) = T(p)$.

Theorem 1. Let $\mathfrak{A} = \{(p,q_1,q_2): p \in E, (q_1,q_2) \in T(p)\}$ be an approximation space. Let $V = \cup \{v(p): p \in E\}$. Let $T^*(p)$ be the reflexive order relation defined in V by $T^*(p) = T(p) \cup \{(q_1,q_2), (q_2,q_2): q_1 \in v(p), q_2 \in V \setminus v(p)\}$. Then $\mathfrak{A}^* = \{(p,q_1,q_2): q_1,q_2 \in V \text{ and } (q_1,q_2) \in T^*(p)\}$ is equivalent to $\mathfrak{A}$ in the sense that $A \subseteq V$ is a best approximating set for p in $\mathfrak{A}^*$-space if and only if $A \cap v(p)$ is a best approximating set in $\mathfrak{A}$-space.

Proof. This theorem enables extension of an approximation space so that $v^*(p) \equiv V$ in the extended space $\mathfrak{A}^*$. Note first that $T^*(p)$ adds the minimum number of elements (q_1,q_2) possible to $T(p)$ to maintain transitivity and the property that $v(p)$ is a best approximating set. Since $v(p)$ is assumed not empty, then $(q_1,q_2) \in T^*(p)$ for $q_1 \in v(p)$ and $q_2 \in V \setminus v(p)$ shows that $A \cap v(p)$ is a best approximating set for p in $\mathfrak{A}$-space if and only if A is a best one in $\mathfrak{A}^*$-space. Obviously $T^*(p)$ is a relatively reflexive order relation. : :

In general, interest is centered on best approximating sets which are somehow minimal. In view of this fact I observe that two approximation spaces in E may be also considered equivalent provided A_1, A_2 are best approximating sets for $p \epsilon E$ respectively in spaces $\mathfrak{A}_1$ and $\mathfrak{A}_2$; then $A_1 \cap \nu_2(p)$ is a best approximating set for p in $\mathfrak{A}_2$-space and $A_2 \cap \nu_1(p)$ is a best approximating set for p in $\mathfrak{A}_1$-space. In particular then, $\nu_1(p) \cap \nu_2(p)$ must be a best approximating set for p in both $\mathfrak{A}_1$ and $\mathfrak{A}_2$-spaces.

Theorem 2. Let $\mathfrak{A} = \{(p,q_1,q_2)\colon p \epsilon E,\ q_1,q_2 \epsilon \nu(p),\ (q_1,q_2) \epsilon T(p)\}$ be an approximation space. Then to each set $A \subseteq V$ which is a best approximating set for p there exists at least one subset A_0 of A which has minimum cardinal number and which is a best approximating set for p. If A_0 is a finite set (in particular if $\nu(p)$ is finite) then A_0 is a minimal best approximating set for p. If there is a unique best approximant q_0 of p then $q_0 \epsilon A$ for every best approximating set A.

Proof. Let $\{A_\alpha\}$ $\alpha \geq 1$, $A_1 = A$ be a complete descending chain of best approximating sets for p, i.e., $A_\alpha \subseteq A_\beta$ if $\alpha > \beta$. Let $\mathfrak{n} = \inf \{|A_\alpha|\}$ where $|A_\alpha|$ is the cardinal number of A_α. Then let A_0 be any set A_α such that $|A_\alpha| = \mathfrak{n}$. Then A_0 is a best approximating set for p and has minimum cardinal, i.e., no proper subset of A_0 of smaller cardinal is a best approximating set for p. If $|A_0|$ is finite then, naturally A_0 is a minimal best approximating set for p. : :

Lemma 3. If $q_0 \epsilon \nu(p)$ is a best approximant of p, then the set A of all elements which are best approximants to p is

$$\{q\colon (q,q_0) \epsilon T(p)\}.$$

Proof. Since $(q_0,q) \epsilon T(p)$ for every $q \epsilon \nu(p)$ when q_0 is a best approximant of p, it follows that $q \neq q_0$ is also a best approximant of p if and only if $(q,q_0) \epsilon T(p)$ from which since $T(p)$ is transitive, $(q,q_1) \epsilon T(p)$ for every $q_1 \epsilon \nu(p)$. : :

In my estimate, it is of little interest, theoretically, whether or not a best approximant q_0 of p exists or is unique. Clearly, the following lemma holds.

Lemma 4. There exists a best approximant q_0 of p if and only if $\nu(p)$ has at least one minimal element. If $T(p)$ is also antisymmetric, then a best approximant of p is unique, if it exists.

Note. In ordinary applications $T(p)$ will not be antisymmetric even when unique best approximants exist. If I let $u(q) = \{q_1\colon (q_1,q) \epsilon T(p)\}$ for each $q \epsilon \nu(p)$, then a necessary and sufficient condition that a unique best approximant q_0 of p exist is that $\{q_0\} = \cap \{u(q)\colon q \epsilon \nu(p)\}$.

Examples

1. Let $E\{\mathrm{f}_1, \cdots, f_n, \cdots\}$ be the set of monomials $f_n(t) \equiv t^n$ limited to

$[-1,1]$. Let $v(f_n)$ be the set of all polynomials of degree at most $n-1$. Let $(q_1,q_2) \epsilon T(f_n)$ provided $||q_1-f_n|| \leq || q_2-f_n||$ where $|| \cdot ||$ is a suitable norm, e.g., the Tchebycheff norm. With this latter choice of norm there exists, as is well known, a unique best approximant of f_n for each n—a Tchebycheff polynomial. For other choices of norm there may be a set of best approximants. Note here that $v(f_n)$ never contains f_n and in general, approximation theorists will rarely treat problems in which the approximated object p is known to be in $v(p)$.

2. Let E be the set of real numbers. Let $v(p) \equiv V$, the set of all integers. Let $(q_1,q_2) \epsilon T(p)$ provided $|q_1-p| \leq |q_2-p|$. Now $T(p)$ is transitive and reflexive and, if $p \epsilon V \subseteq E$, then p is the best approximant of itself. There exists for every p not of form $q+\frac{1}{2}$, for $q \epsilon V$ a unique best approximant which is the integer "closest" to p. If $p=q+\frac{1}{2}$ for $q \epsilon V$, then $\{q, q+1\}$ are equally best approximants of p.

3. Let E be the set of real-valued functions which are of class $C^{(1)}$ on the open interval $(-1,1)$ and continuous on $[-1,1]$. Let $v(p) \equiv V$ be the linear space of polynomials of degree at most n. Let $(q_1,q_2) \epsilon T(p)$ provided $|| q_1-p || \leq || q_2-p ||$ *and* $|| q_1'-p' || \leq || q_2'-p' ||$ where $|| \cdot ||$ is a suitable norm. Now there will be best approximating sets of polynomials for a given p, but in general, no best approximant, unless $p \epsilon v(p)$.

4. Let E be the set of real numbers. Let $v(p)=V$ be the set of all integers. Let $(q_1,q_2) \epsilon T(p)$ provided $\sin(q_1-p) \leq \sin(q_2-p)$. Now $T(p)$ is reflexive and transitive and there will exist, in general, a sequence $\{q_n\}$ which is a best approximating set for p but ordinarily not a best approximant unless, for example, p is of the form $\pi/2+2n\pi$, n an integer. Note that even if $p \epsilon v(p)$ here, p may not be a best approximant to p. For example, if $p=0$, $q=0$ is not a best approximant since $\sin 4<0$ and $\sin 5<0$.

The last example suggests that criteria for approximation more general than metrics and norms can well be used. For convenience I will limit myself to a real-valued function, d, mapping $E \otimes V$ into the set of real numbers. Then I define $(q_1,q_2) \epsilon T(p)$ provided $d(p,q_1) \leq d(p,q_2)$.

Theorem 5. Let $\mathfrak{A} = \{(p,q_1,q_2): p \epsilon E, d(p,q_1) \leq d(p,q_2), q_1,q_2 \epsilon V\}$ where E and V are not empty. Then to each $p \epsilon E$ there exists a sequence $\{q_n\}$ which is a best approximating set for p where the q_n are not necessarily distinct.

Proof. Let $S(p) = \{d(p,q): q \epsilon V\}$. Then $S(p)$ is a set of real numbers and hence there exists a sequence $q_1, \cdots, q_n, \cdots$ of elements of V such that $d(p,q_n) \leq d(p,q_{n-1})$ $n \geq 2$ and if $r \epsilon S(p)$ then for some positive integer m $d(p,q_n) \leq r$ for $n \geq m$. Hence $\{q_n\}$ is a best approximating set for p. : :

Remarks

It should be observed that best approximating sequences exist here simply because of the order property of real numbers, not because of any other extraneous properties attributed to the functions d, such as being nonnegative, symmetric, or satisfying the triangle law. In many approximation problems the existence of such a sequence is more important than the existence of a best approximant. It should be noted that if $T(p)$ is a reflexive order relation then so is $T^{-1}(p) = \{(q_2,q_1): (q_1,q_2) \in T(p)\}$ and hence "worst" approximations are also "best" when $T^{-1}(p)$ is the criterion.

III. ISOTONIC SPACES AND APPROXIMATION SPACES

It is clear that the neighborhood concept of topology is intimately connected to the notion of closeness, i.e., approximation. There is a significant generalization of topological spaces which I call isotonic spaces. The most general form of these is presented here.

Let now E be a set and set D be another set not necessarily related to E. Let $V = PD$, the power set of D, i.e., the class of all subsets of D. Let $v(p) \equiv \mathscr{V}(p)$ an ancestrally closed subclass of V. If $Y \in \mathscr{V}(p)$, then Y is called a neighborhood of p and $\mathscr{V}(p)$ is the class of all neighborhoods of p. Note that a neighborhood of p is a subset of D, not of E. Let $\mathfrak{C}(p) = \{X: X \subseteq D, X \cap Y \text{ is not empty for all } Y \in \mathscr{V}(p)\}$. Then if $X \in \mathfrak{C}(p)$, X is called a *convergent* of p. Now $\mathfrak{C}(p)$ is also an ancestrally closed subclass of V. The collection $\{\mathscr{V}(p): p \in E\}$ is called an *isotonic space* in (E,D).

Theorem 6. Let $\mathfrak{A} = \{(p, Y_1, Y_2): p \in E,\ Y_1, Y_2 \in \mathscr{V}(p),\ Y_1 \subseteq Y_2\}$. Then the approximation space $\mathfrak{A}$ is equivalent to the isotonic space $\{\mathscr{V}(p): p \in E\}$ and a subclass $\mathscr{V}_0(p)$ of $\mathscr{V}(p)$ is a best approximant of p if and only if $\mathscr{V}(p)$ is the ancestral closure of $\mathscr{V}_0(p)$, i.e., $\mathscr{V}_0(p)$ is a neighborhood base for p. : :

Proof. Suppose $\mathscr{V}_0(p)$ is a best approximant of p. Then to each $Y \in \mathscr{V}(p)$ there must be $Y_0 \in \mathscr{V}_0(p)$ such that $Y_0 \subseteq Y$. Hence $\mathscr{V}_0(p)$ is a base for $\mathscr{V}(p)$ necessarily. Obviously if $\mathscr{V}_0(p)$ is a base for $\mathscr{V}(p)$ then $\mathscr{V}_0(p)$ is a best approximant of p. : :

Comment

It is now readily seen why it is simpler to follow H. Cartan and use the neighborhood filter for p (generalized) than to use the convergents.

If I identify $D=E$ and $V=PE$ and if I take $\mathscr{V}(p)$ to be the filter of all neighborhoods of p in the topological sense, then the resulting approximation space is equivalent to the topological space in E. Hence all isotonic and topological spaces are special kinds of approximation spaces.

The difficulty of using generalized nets will now be illustrated. For simplicity let me deal with one point $p \epsilon E$ and hence one subclass $\mathscr{V}(p)$ of $V=PD$ which is a neighborhood class for p. Let s be a function mapping $\mathscr{V}(p)$ into D so that to each neighborhood Y is assigned a unique element sY of Y. How should I generate an order relation in the elements sY? The simple answer is that I will treat sY like a convergent sequence $\{a_n\}$, and write $sY_1 \leq sY_2$ provided $Y_1 \subseteq Y_2$. That is, it is not the elements of D which are ordered but the "subscripts." Note now that the set $X=\{sY: Y \epsilon \mathscr{V}(p)\}$ is necessarily a convergent of p since $sY \epsilon Y$. Now let s_i range over all functions which select one element from each $Y \epsilon \mathscr{V}(p)$ for each $i \epsilon I$ an index set. Then I define $\mathfrak{A}=\{(p,[Y_1, \{s_i\}],[Y_2,\{s_i\}]): p \epsilon E,\ Y_1, Y_2 \epsilon \mathscr{V}(p),\ Y_1 \subseteq Y_2\}$ and this is an approximation space in which, if $\mathscr{V}_0(p)$ is a base of $\mathscr{V}(p)$, $X_i=\{s_i Y: Y \epsilon \mathscr{V}_0(p)\}$ is a convergent of p and $\{X_i: i \epsilon I\}$ for any such $\mathscr{V}_0(p)$ is a base for $\mathfrak{C}(p)$. The point is, that to bring in convergence theory the order relation $T(p)$, in general, must reflect the inclusion relation in the neighborhood class, and cannot simply be an order relation among elements of D. For this reason it is simplest to deal with the neighborhood filter directly. In view of this negative outlook, the following theorem is of interest.

Theorem 7. Let $v(p)=E\setminus\{p\}$ and let $T(p)$ be a reflexive order relation which directs $E\setminus\{p\}$ for each $p \epsilon E$. That is, to each $q_1,q_2 \epsilon v(p)$ there exists $q_3 \epsilon v(p)$ such that (q_3,q_1), $(q_3,q_2) \epsilon T(p)$. Let, for each $q \epsilon E\setminus\{p\}$ $u(q)=\{q_1: (q_1,q) \epsilon T(p)\}$. Then the class $\mathscr{V}_0(p)=\{\{p\} \cup u(q): q \epsilon v(p)\}$ is a base for a class of topological neighborhoods of p. If $\mathfrak{A}=\{(p,q_1,q_2): p \epsilon E,\ (q_1,q_2) \epsilon T(p)\}$, then a subset A of E is a best approximant of p if and only if p is a limit (accumulation) point of A. Thus $\mathfrak{A}$ is equivalent to a topological space.

Proof. Since $T(p)$ is a reflexive order relation in $E\setminus\{p\}$, it is necessary only to show that $\mathscr{V}_0(p)$ is a base of topological neighborhoods to show that if $Y_1,\ Y_2 \epsilon \mathscr{V}_0(p)$ then there is $Y_3 \epsilon \mathscr{V}_0(p)$ such that $Y_3 \subseteq Y_1 \cap Y_2$, since $p \epsilon Y$ if $Y \epsilon \mathscr{V}_0(p)$. Now if $Y_1=\{p\} \cup u(q_1)$ and $u(q_1)=\emptyset$ then $Y_1=\{p\}$, $Y_1 \cap Y_2=\{p\}$ and certainly $Y_3=\{p\}$ satisfies the requirement. Hence suppose $u(q_1)\neq\emptyset$ and $u(q_2)\neq\emptyset$, where $Y_i=\{p\} \cup u(q_i)$; $i=1,2$. Let $q_3 \epsilon v(p)$ as required by the directedness condition in relation to q_1, q_2. Then $u(q_3) \subseteq u(q_1) \cap u(q_2)$ and hence $Y_3=\{p\} \cup u(q_3) \subseteq Y_1 \cap Y_2$. Hence the class $\mathscr{V}_0(p)$ is a neighborhood base for p in the topological sense. The rest of the theorem follows directly. : :

This theorem then shows that many popular topological spaces such as metric spaces are simply representable by a ternary relation in E which is an approximation space. The reader should notice that the topologies indicated by Theorem 7 are not the kind referred to as order topologies. In an order topology a fixed antisymmetric reflexive order relation provides the neighborhoods for all points. Here I am dealing with reflexive transitive relations associated with each point. This framework is more general but different in approach. It is possible to obtain any topology in E by taking collections of approximation spaces of the sort described in Theorem 7, but this is rather an unattractive prospect, in view of the simplicity of using neighborhood filters.

IV. EXAMPLE OF AN APPERT SPACE

The reader may feel that isotonic spaces which are not topological are rare items. This is far from being the case. Every closure under any set of operations provides a special kind of isotonic space which is called an Appert space. Rather few Appert spaces are topological spaces. As an example let E be the plane and let $\mathcal{V}(p)$ be the class of all subsets of E which contain some coconvex set of which p is an element. A coconvex set is the complement of a convex set. In this Appert space the convex sets are the closed sets and their complements are the open sets. Now $\mathfrak{C}(p)$, the class of convergents of p, is the class of all sets X which have p in their convex hulls. The approximation space $\mathfrak{A} = \{(p, Y_1, Y_2): p \in E, Y_1, Y_2 \in \mathcal{V}(p), Y_1 \subseteq Y_2\}$ is equivalent to the Appert space. In this case there exists a unique minimal neighborhood base for p—namely, the set of all cosemispaces at p where a semispace at p I define to be a maximal convex subset of E which excludes p. The cardinal number of the neighborhood base is the same as that of the real number set. This Appert space has no representation as an approximation space which is a ternary relation in E.

V. UNIFORM APPROXIMATION SPACES

Uniform approximation spaces are here introduced as a generalization of approximation spaces of real-valued functions. First I summarize a few results proved by Sister Gregory Michaud and myself (1967) concerning order relations. Let T be an order relation in a set M. Then $T^{-1} = \{(q,p): (p,q) \in T\}$. The set $S_0 = T \cap T^{-1} \subseteq M \otimes M$ is the symmetric part of T and the set $S_1 = T \setminus S_0$ is the asymmetric part of T. Either may be empty but both are transitive, i.e., order relations. Suppose (p,q),

$(q,p) \notin T$ and $p \neq q$. Then there exists a maximal order relation T_1, say, which contains T (as a subset of $M \otimes M$) such that $(p,q) \in T_1$, and $T_1 \cap T_1^{-1} = T \cap T^{-1} = S_0$. Hence T is the intersection of a family of such maximal order relations (i.e., transitive relations). Note that T_1 is a total relation, i.e., if $p \neq q$ then (p,q) or $(q,p) \in T$.

Suppose T is reflexive. Then each maximal order relation T_1 such that T_1 contains T and $T_1 \cap T_1^{-1} = S_0$, has the property that T_1 is total and reflexive. Then T_1 induces a linear ordering of the equivalence sets in M determined by S_0. Let R be a set which has the same cardinal number as the set of equivalence classes determined by S_0. Let d map each equivalence class into one element of R so that d is onto. Let $\leq$ be a linear order in R determined by T_1, i.e., $dp \leq dq$ if and only if $(p,q) \in T_1$. Then $dp = dq$ if and only if $(p,q), (q,p) \in S_0 \subseteq T_1$. If desired R, $\leq$ may be embedded isomorphically in a linear order R^*, $\leq$ which has the Dedekind cut property and which has a maximum and a minimum element. If $r_1 < r_2$ in R and there is no r in R such that $r_1 < r < r_2$ then a segment isomorphic to the real number order may be inserted between r_1 and r_2. I propose to use such a linear order R or an extension R^* as a radius set.

Now let E, V be two sets and let T be a reflexive order relation in $E \otimes V$, i.e., T is an order relation in the set of ordered pairs (p,q), $p \in E$, $q \in E$. Then I interpret T as follows. The statement $[(p_1,q_1),(p_2,q_2)] \in T$ shall mean "q_1 is at least as good an approximant of p_1 as q_2 is of p_2." Then T is called a uniform approximation space in $E \otimes V$ provided it is also a total relation, i.e., (p_1,q_1) and $(p_2,q_2) \in E \otimes V$ $(p_1,q_1) \neq (p_2,q_2)$ implies $[(p_1,q_1),(p_2,q_2)] \in T$ or $[(p_2,q_2),(p_1,q_1)] \in T$ (or both). Now let $T(p)$ for $p \in E$ be defined as follows: $T(p) = \{(q_1,q_2): [(p,q_1),(p,q_2)] \in T\}$. Since $T(p)$ is a reflexive order relation in V (which is also total) I have:

Lemma 8. Every uniform approximation space T in $E \otimes V$ induces an approximation space $\mathfrak{A} = \{(p,q_1,q_2): p \in E, q_1,q_2 \in V \text{ and } (q_1,q_2) \in T(p)\}$ where $T(p)$ is as defined above.

Now let $R, \leq$ be a linear order such that there exists a function d mapping $E \otimes V$ into R and $d(p_1,q_1) \leq d(p_2,q_2)$ provided $[(p_1,q_1),(p_2,q_2)] \in T$ where T is a uniform approximation space in $E \otimes V$. Let $p \in E$, $r \in R$. Let $S(p,r) = \{q: q \in V, d(p,q) \leq r\}$ be called a closed sphere with center p and radius r. The following result suggests itself.

Theorem 9. Let d be a mapping of $E \otimes V$ into a linear order $R, \leq$. Then there exists a unique uniform approximation space T in $E \otimes V$ defined by $T = \{(p_1,q_1),(p_2,q_2)]: d(p_1,q_1) \leq d(p_2,q_2)\}$. Hence each uniform approximation space in $E \otimes V$ is equivalent to some mapping of $E \otimes V$ into some linear order $R, \leq$ and, in this sense, every uniform approximation space is a "metric" space.

Remarks

Since the linear order R is assumed reflexive and since d maps $E \otimes V$ into R, T as defined above is a reflexive total order relation in $E \otimes V$ and hence is a uniform approximation space. The advantages of uniform approximation spaces are numerous. To capitalize on some of them, it is convenient to assume that the linear order $R, \leq$ is order complete and has a minimum element 0 and a maximum element 1. What is usually done is to use some particular set R such as the set of nonnegative real numbers where the maximum element is labeled $+\infty$. In general one may find that such an R is inadequate. The cardinal number of R given E, V need not be greater than the cardinal number of $E \otimes V$ for a given T.

In a uniform approximation space T comparisons of the approximation of a point by sets and of sets by sets may be made in several ways. For example, I might define $d(A,B)$ for $A \subseteq E$, $B \subseteq V$ by $d(A,B) = \{d(p,q) : p \in A, q \in B\}$ which is a subset of R. Now if R is order complete with maximum and minimum elements, then $d_0(A,B) = \inf d(A,B) \in R$ is the minimum "distance" from A to B and $d_1(A,B) = \sup d(A,B)$ is the maximum distance. Another approach is to define $d^*(A,B) = \sup_{p \in A} \{\inf_{q \in B} \{d(p,q)\}\}$. If $\inf_{q \in B} \{d(p,q)\} = r_1 \in R$ then $S(p,r)$ is a sphere such that if $r_1 < r$ then there exists $q \in B$ such that $d(p,q) < r$ but no $q \in B$ exists such that $d(p,q) < r_1$. Thus $d^*(A,B)$, if achieved by $p_0 \in A$, $q_0 \in B$, would be the radius of a largest radius sphere with center in A and no point of B in its "interior." Now if $B_1 \supset B$, then $d^*(A,B_1) \leq d\ (A,B)$ but if $A_1 \supset A$ then $d^*(A_1,B) \geq d^*(A,B)$, i.e., it is "harder" to approximate a larger set by a given set but a larger set of approximants is "closer" to a given set of objects to be approximated.

Theorem 10. Let T be a uniform approximation space in $E \otimes V$. Then a subset A is a best approximating set for $p \in E$ if and only if $A \cap S(p,r)$ is not empty for every $r \in R$ for which $S(p,r)$ is not empty.

Proof. Since A is a best approximating set for p if and only if to each $q \in V$ there is a $q_0 \in A$, with $(q_0,q) \in T(p)$ and this is equivalent to $d(p,q_0) \leq d(p,q)$, it follows that with $r = d(p,q)$ that $q_0 \in S(p,r)$ and hence A intersects every nonempty $S(p,r)$. Again if A is a subset of V which intersects every nonempty $S(p,r)$ then with $r = d(p,q)$ for $q \in V, q \in S\ (p,r)$ and hence $A \cap S(p,r)$ has an element q_0, say. But then $(q_0,q) \in T(p)$ since $d(p,q_0) \leq d(p,q)$. Hence A is a best approximant of p. : :

Corollary 11. Let $A \subseteq V$ be a best approximant of $p \in E$ in a uniform approximation space T in $E \otimes V$. Then there exists a subset A_0 of A

which is strictly linearly ordered by $T(p)$ and which is a best approximant of p.

Proof. The set $R_p = \{d(p,q): q \epsilon A\}$ is a subset of R and since R is linearly ordered, let us pick for each $r \epsilon R_p$ one element q of A such that $d(p,q)=r$, and let the resulting set be A_0. Then if $(q_1,q_2) \epsilon A_0$, it follows that $q_1 \neq q_2$ and $d(p,q_1) < d(p,q_2)$ or $d(p,q_2) < d(p,q_1)$ but not both. Hence A_0 has a strict linear order imposed by $T(p)$ and A_0 is a best approximant of p by Theorem 10. : :

Remarks

It is clear that a uniform approximation space is intuitively very close to being a metric space. A "neighborhood base" for p is at hand which is linearly ordered by inclusion and thus the image of convergence in the usual sense is present. Now the linear order $R, \leq$, as I have pointed out, in effect, can be any one which produces the uniform approximation space. Of course, if I can use a set of real numbers, then a best approximating set A_0 can be a sequence in the usual sense.

One way of generating uniform approximation spaces may be mentioned. To each $p \epsilon E$ let there be associated a class $\mathcal{V}_0(p) = \{Y_r\}$ of subsets of V where $r \epsilon R$ a set linearly ordered by $\leq$, such that $r_1 < r_2$ implies $Y_{r_1} \subseteq Y_{r_2}$ where R is the same for all p. Let R be order complete and have a minimum 0 and a maximum element 1. Let $d(p,q) = \inf \{r: q \epsilon Y_r,\ Y_r \epsilon \mathcal{V}(p)\}$. If $q \notin Y_r$ for any $Y_r \epsilon \mathcal{V}(p)$ then define $d(p,q)=1$. The resulting map $d: E \otimes V \rightarrow R$ gives a uniform approximation space T.

Before leaving uniform approximation spaces, let me introduce a generalization. I assumed before that the uniform approximation space should involve a total order relation, i.e., that every pair (p_1,q_1), (p_2,q_2) should be represented. The reason for doing this was that the word uniform suggested to me that one would like to compare every pair (p_1,q_1), (p_2,q_2). However, by dropping the requirement that T be total, a larger variety of examples may be directly reached.

Theorem 12. Let T be a reflexive order relation in $E \otimes V$. Let $\{T_i: i \epsilon I\}$ be family order relations in $E \otimes V$ the intersection of which is T. Let T_i be equivalent to a mapping d_i of $E \otimes V$ into a linear order $R_i, \leq$ for $i \epsilon I$. Then $[(p_1,q_1),(p_2,q_2)] \epsilon T$ if and only if $d_i(p_1,q_1) \leq d_i(p_2,q_2)$ for every $i \epsilon I$. If $A \subseteq V$ is a best approximant of p in every T_i-space, then A is a best approximant of p relative to T, which is a best approximant of p relative to T.

VI. CONCLUSION

It may appear to the uninitiated that there is little to be done with the approximation spaces introduced here. This is true on one count; namely the general theory does little to solve the problems of approximation; it raises instead many further problems. However, the general theory has several features which make it worthwhile. First it shows that topological and isotonic spaces are instances of a more general concept of approximation spaces and this enables and suggests using some of the numerous results in topology to suggest lines of inquiry into the theory of approximation spaces. The general theory pinpoints the meaning of approximation as specialized ones have been unable to do. The deliberate use of a set $v(p)$ not in the same set E as p allows inclusion of computer models, simulations and so on, as approximants. The assumption of transitive (i.e., order) relations brings up the difficulties of nonformal approximation—it is not easy to establish criteria of goodness of fit in many cases.

However, that is not all. If you like to work on particular and difficult problems you can generate them in quantity. For example, just which properties of a function should an approximation to it preserve? Its extremes? The sign of its derivative? The direction at infinity? The integral of its square? Its zeroes?

How do you compare two finite difference schemes for solving a differential equation? Two algorithms for estimating eigenvalues of a matrix? Two models of the same system? Obversely, how well does a constructed mechanism approximate a formal one?

The study of approximation and hence of optimization is germane to the activities of everyone. It is not well embedded in normed or metric spaces!

ACKNOWLEDGMENT

Supported partially by the United States National Science Foundation Grant for Computer Science and Logic, Grant Number GP-7077.

REFERENCES

Gastl, George and P. C. Hammer. 1967. "Extended Topology: Neighborhoods and Convergents," *Proc. Coll. on Convexity,* Copenhagen, 1965, pp. 104–116.

Gregory Michaud, Sister and P. C. Hammer. 1967. "Extended Topology: Transitive Relations," *Nieuw Archief voor Wiskunde* Ser. 37 **15**: 38–48.

Hammer, P. C. 1963. "Topologies of Approximation," *SIAM Numerial Analysis J.* **1**: 69–75.

Hammer, P. C. 1965. "Extended Topology: Caratheodory's Theorem on Convex Sets," *Rend. Circ. Mat. Palermo* Ser. 2 **11**:34–42.

Hammer. P. C. 1966. "Extended Topology and Systems," *Math. Systems Theory J.* **1**: 135–142.
Hammer, P. C. 1967a. "Isotonic Spaces in Convexity," *Proc. Coll. on Convexity,* Copenhagen, 1965, pp. 132–141.
Hammer, P. C. 1967b. "Generalization in Geometry Lectures I and II. Notes by M. Hausner," Report on CUPM Conference on Geometry Part III. Berkeley, Calif.
Hammer, P. C. 1967c. "Language, Approximation and Topologies," *Advances in Linguistics.* U. Wisconsin Press, Madison, Wis. pp. 33–40.

These references provide some of the flavor of the general theory of spaces (extended topologies) which are germane to approximation more or less directly. Gastl and Hammer (1967) contains a discussion of isotonic spaces useful in convexity. Hammer (1963) contains detailed examples omitted here. Hammer (1966) shows a connection between systems theory and isotonic spaces. Hammer (1965) shows how a new set of results were generated using Appert spaces. This led to the use of transitive relations in Hammer (1967a and b) to generalize convexity.

7

CHART OF ELEMENTAL MATHEMATICS

PRESTON C. HAMMER

I. WHAT IS MATHEMATICS?

During the past 20 years I have been engaged in many activities which could scarcely be called mathematical, in positions which I held largely because I was considered a mathematician. One of the perpetual frustrations mathematicians face in trying to apply their prior knowledge is the inadequacy of the concepts, techniques, and facts to provide solutions to the problems in hand. To find as I did, for example, that solving a quadratic equation numerically requires finesse, that problems of sorting were far from trivial, and that error analyses were not useful for many real problems, were enlightening but not comforting experiences. Even greater discomfort arose when I found that the computing and approximation business into which I drifted could not make effective use of mathematical concepts which, I had been told, were very important.

Thus the concepts of continuity, compactness, measure, neighborhood, limit, and so on got almost left out in the finitistic world of computing and language. Having been miseducated to believe that mathematics was general, so general that much of it applied, if at all, only to mathematics, I had to spend more than a reasonable amount of energy in learning some facts. Gradually some light dawned. Many fields of mathematics, touted as being far out in terms of generality, began to fall into place as specialized areas insufficiently general to embrace instances known to school children. I found that areas of algebra such as lattices, semigroups, and groupoids were useful but not adequately general. Topological spaces often considered in their general form to be much too general by many, turned out to be embraced in a more general and more applicable space theory.

These items had been realized by several others. My work, however, differed from those with the leisure to pursue their mathematics without interruption. I urge generality as a practical principle to be used as a means of interpreting the swelling flood of mathematical information. Mathematics education today is comparable to the slow advances enforced on apprentices by guilds and trade unions of a former era. This is evidently due to several factors which may be discerned as the failure of mathematicians to understand mathematics and the natural resistance to change in one of the most conservative fields there is.

What do I mean? I mean that an undergraduate mathematics major, having had mathematics in school through college, is still left with a false impression of mathematics. He will have no sense of mathematics as a part of the culture, he will not know that it is not complete, that new areas beckon him for work and old ones for improvements. Why should this be so? The only answer possible is miseducation. The student is being *trained.* He is acquiring comparatively disorganized facts and skills but he sees no pattern. Mathematicians are so busy doing things that they take little time to think! Yet thinking is required in order to understand mathematics.

How could thinking people have become aware of order relations explicitly only around 1900 when they occur everywhere in the common language (comparative adjectives), and everywhere in mathematics? Who but a nonthinking mathematician would warn others against studying order relations except for their application? Semigroups are still mainly reserved for graduate school, although they are needed and present from elementary arithmetic on. How many years have gone by since Cauchy introduced the word "continuity" into mathematics? Why has no one before me stated the general sense of continuity, so well concealed in the mathematical treatments? The answer lies in the failure to think.

The time has come when we should recognize that we will not go too far even with the best of our abilities. The life span of man is not so great that we can afford to drain his energy by miseducation. In mathematics, in particular, there is no good reason that a high school graduate should not have some grasp of what mathematics is about.

This essay presents one of the many possible attacks on mathematical illiteracy. This takes the form of a chart of elemental mathematics with some discussion of its plan. No one who has seen this chart has been extremely pleased with its format. This is as it should be. It is an initial

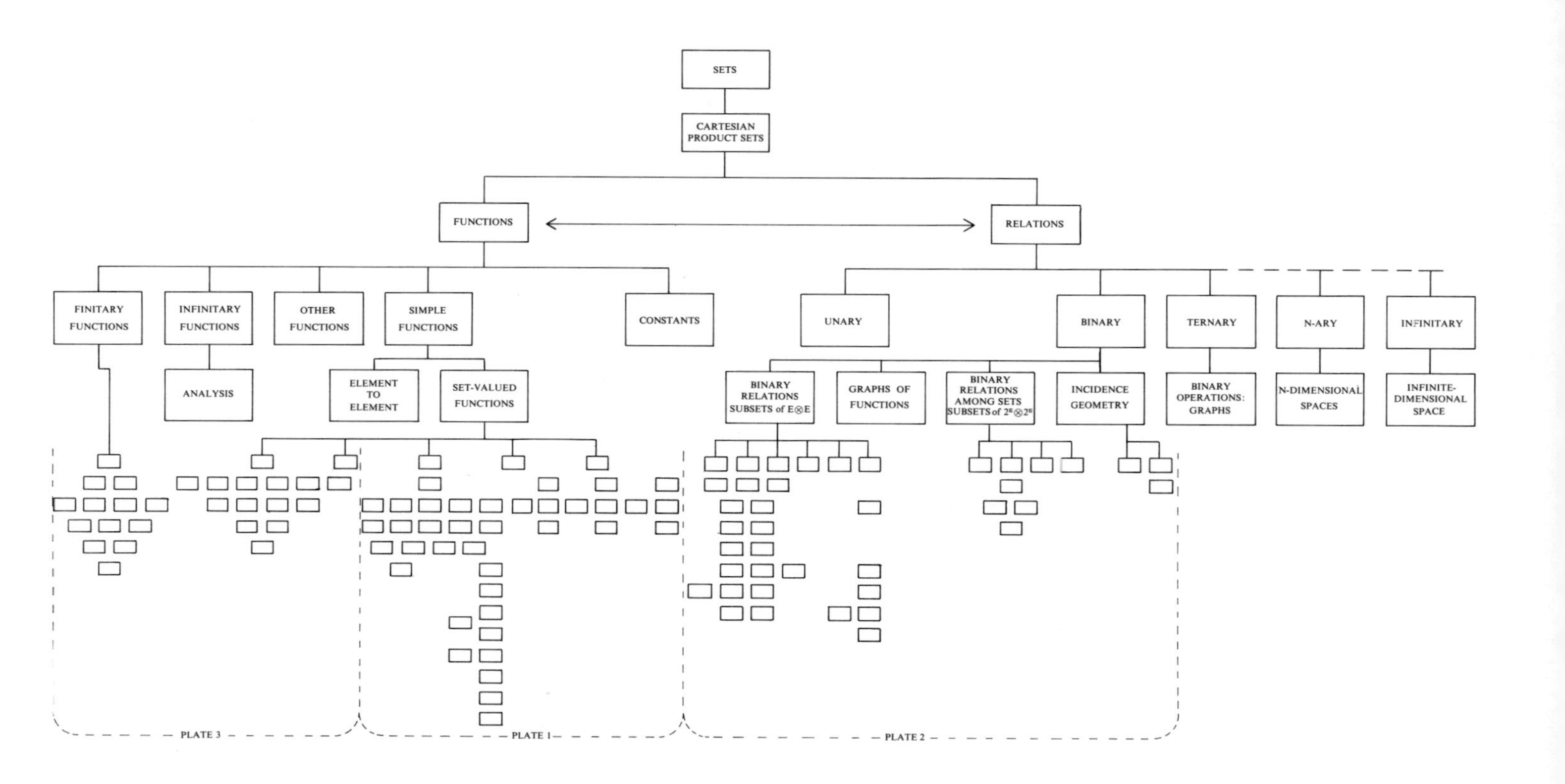

SETS
CARTESIAN PRODUCT SETS
FUNCTIONS
RELATIONS
FINITARY FUNCTIONS
INFINITARY FUNCTIONS
OTHER FUNCTIONS
SIMPLE FUNCTIONS
CONSTANTS
UNARY
BINARY
TERNARY
N-ARY
INFINITARY
ANALYSIS
ELEMENT TO ELEMENT
SET-VALUED FUNCTIONS
BINARY RELATIONS SUBSETS of E⊗E
GRAPHS OF FUNCTIONS
BINARY RELATIONS AMONG SETS SUBSETS of $2^E \otimes 2^E$
INCIDENCE GEOMETRY
BINARY OPERATIONS: GRAPHS
N-DIMENSIONAL SPACES
INFINITE-DIMENSIONAL SPACE
PLATE 3
PLATE 1
PLATE 2

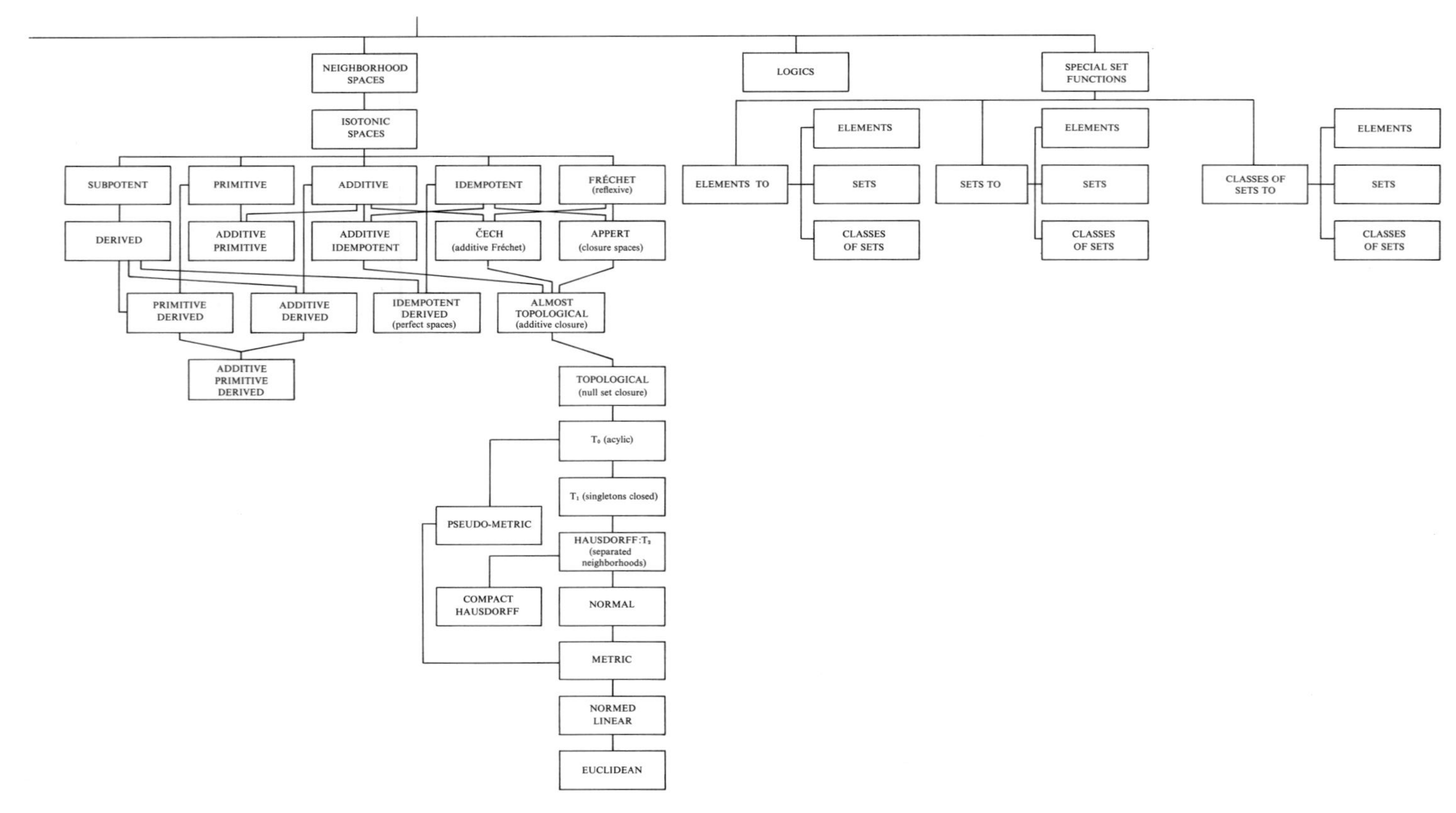
NEIGHBORHOOD SPACES
LOGICS
SPECIAL SET FUNCTIONS
ISOTONIC SPACES
ELEMENTS TO
ELEMENTS
SETS
CLASSES OF SETS
SETS TO
ELEMENTS
SETS
CLASSES OF SETS
CLASSES OF SETS TO
ELEMENTS
SETS
CLASSES OF SETS
SUBPOTENT
PRIMITIVE
ADDITIVE
IDEMPOTENT
FRÉCHET (reflexive)
DERIVED
ADDITIVE PRIMITIVE
ADDITIVE IDEMPOTENT
ČECH (additive Fréchet)
APPERT (closure spaces)
PRIMITIVE DERIVED
ADDITIVE DERIVED
IDEMPOTENT DERIVED (perfect spaces)
ALMOST TOPOLOGICAL (additive closure)
ADDITIVE PRIMITIVE DERIVED
TOPOLOGICAL (null set closure)
T_0 (acylic)
T_1 (singletons closed)
PSEUDO-METRIC
HAUSDORFF:T_2 (separated neighborhoods)
COMPACT HAUSDORFF
NORMAL
METRIC
NORMED LINEAR
EUCLIDEAN

Plate 1

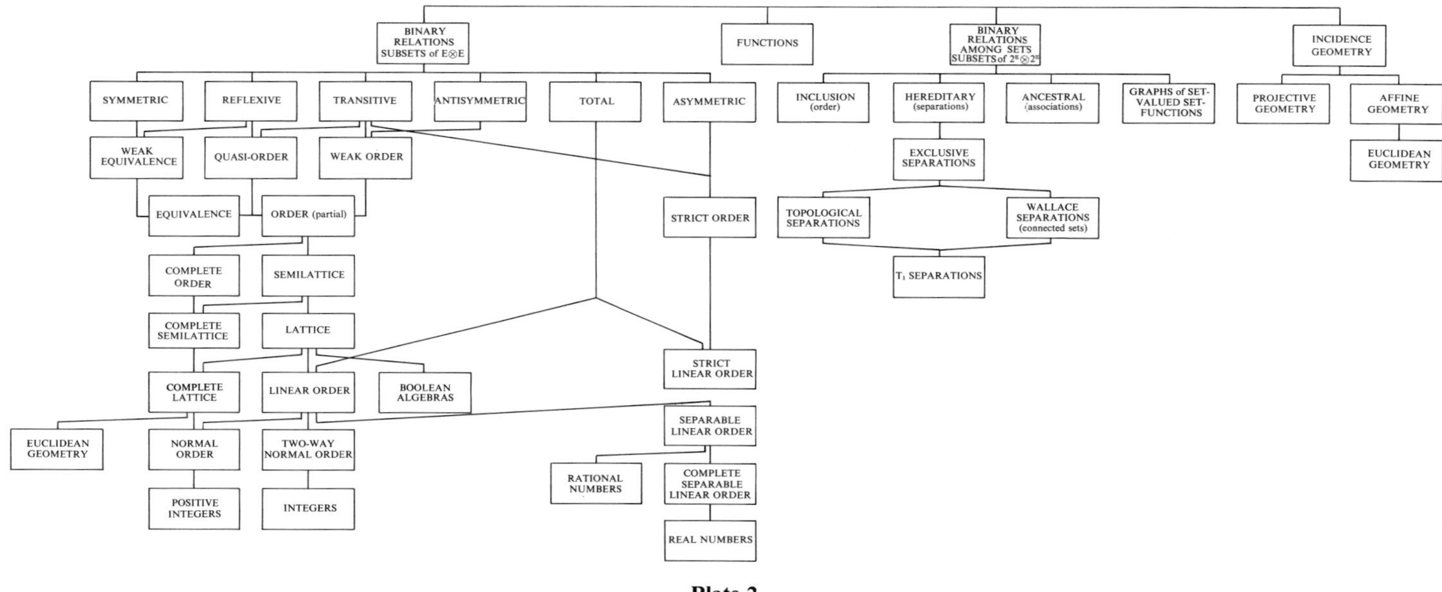
BINARY RELATIONS SUBSETS of E⊗E
FUNCTIONS
BINARY RELATIONS AMONG SETS SUBSETS of $2^E \otimes 2^E$
INCIDENCE GEOMETRY
SYMMETRIC
REFLEXIVE
TRANSITIVE
ANTISYMMETRIC
TOTAL
ASYMMETRIC
INCLUSION (order)
HEREDITARY (separations)
ANCESTRAL (associations)
GRAPHS of SET-VALUED SET-FUNCTIONS
PROJECTIVE GEOMETRY
AFFINE GEOMETRY
WEAK EQUIVALENCE
QUASI-ORDER
WEAK ORDER
EXCLUSIVE SEPARATIONS
EUCLIDEAN GEOMETRY
EQUIVALENCE
ORDER (partial)
STRICT ORDER
TOPOLOGICAL SEPARATIONS
WALLACE SEPARATIONS (connected sets)
COMPLETE ORDER
SEMILATTICE
T_1 SEPARATIONS
COMPLETE SEMILATTICE
LATTICE
COMPLETE LATTICE
LINEAR ORDER
BOOLEAN ALGEBRAS
STRICT LINEAR ORDER
SEPARABLE LINEAR ORDER
EUCLIDEAN GEOMETRY
NORMAL ORDER
TWO-WAY NORMAL ORDER
RATIONAL NUMBERS
COMPLETE SEPARABLE LINEAR ORDER
POSITIVE INTEGERS
INTEGERS
REAL NUMBERS

Plate 2

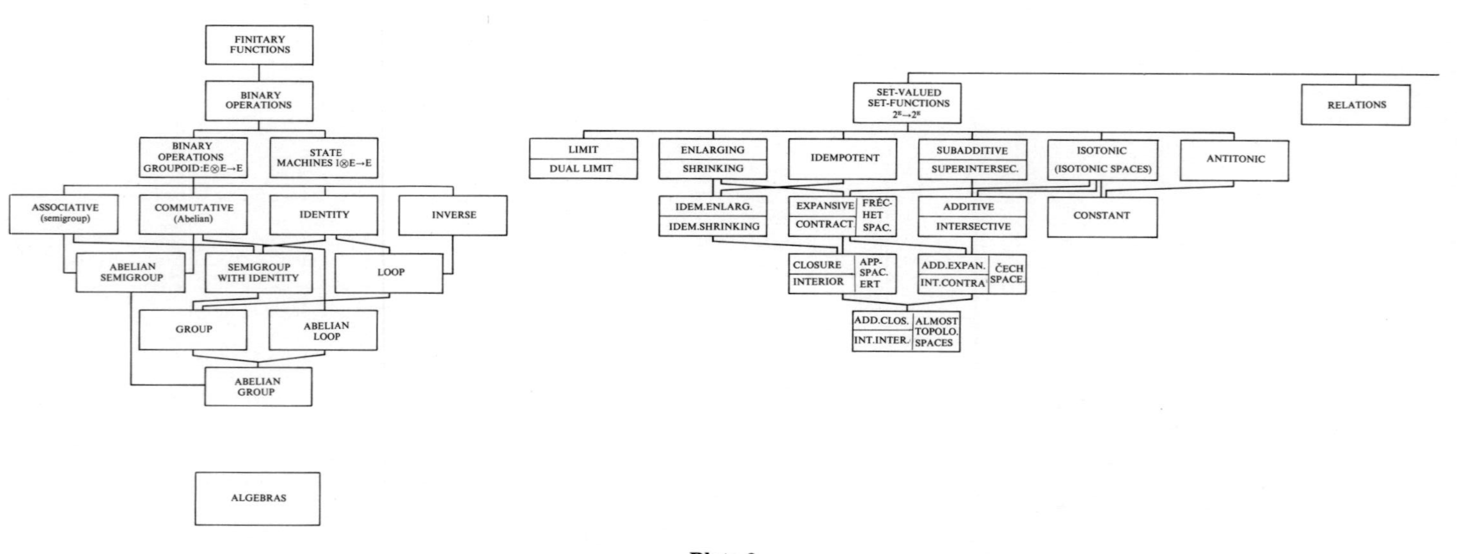
FINITARY FUNCTIONS
BINARY OPERATIONS
BINARY OPERATIONS GROUPOID:E⊗E→E
STATE MACHINES I⊗E→E
ASSOCIATIVE (semigroup)
COMMUTATIVE (Abelian)
IDENTITY
INVERSE
ABELIAN SEMIGROUP
SEMIGROUP WITH IDENTITY
LOOP
GROUP
ABELIAN LOOP
ABELIAN GROUP
ALGEBRAS
SET-VALUED SET-FUNCTIONS $2^E \to 2^E$
RELATIONS
LIMIT
DUAL LIMIT
ENLARGING
SHRINKING
IDEMPOTENT
SUBADDITIVE
SUPERINTERSEC.
ISOTONIC
(ISOTONIC SPACES)
ANTITONIC
IDEM.ENLARG.
IDEM.SHRINKING
EXPANSIVE
CONTRACT.
FRÉC-HET SPAC.
ADDITIVE
INTERSECTIVE
CONSTANT
CLOSURE
INTERIOR
APP-SPAC. ERT
ADD.EXPAN.
INT.CONTRA
ČECH SPACE
ADD.CLOS.
INT.INTER.
ALMOST TOPOLO. SPACES

Plate 3

attempt to get some global pattern into the mathematical structure. As an initial attempt it probably bears less resemblance to what it will become, after improvements, than Mendeleev's periodic table of elements did to its present offspring. Having spent some years in research bearing on global mathematics, I initially tried to draft a chart of mathematical systems in general. This was, frankly, a morass. I then decided to take an analogy from Mendeleev. Why not try, I reasoned, to sort out mathematical systems which are in some way irreducible?

Now, to define irreducibility is not simple and I failed to do it. However, it did seem to me, for example, that a group is scarcely a compound of two "important" systems, whereas a ring or a field involving two binary operations may well be considered as composite. Thus, I decided to search out "atomic" areas from which other systems may be composed. The analogy with the periodic table of elements can be carried only so far. Mendeleev was dealing with objects in his thinking, but mathematics has no objects in that sense, it is a creation of people—a biopsycholinguistic field.

It would not do to simply state that mathematics is a collection of theorems about (abstract) sets and let it rest since this would carry too little information. On the other hand, detailed classification of functions, even on a comparatively simple basis, couldn't be allowed. The classification must be coarse but not too coarse. Thus I decided to include functions and relations, to give a crude classification of each, and to follow this on lower levels to give some idea of how special subjects fit and where certain popular areas appear.

The importance of a mathematical area generally is not reflected in the chart. Arithmetic, elementary algebra, and geometry are the most important branches of mathematics when gauged by their uses. None of these appears. Universal algebra and category theory do not appear because of their superstructural character. They may provide methods of better classification but they do not fit this chart. Cartesian products reflect our simple way of writing. I decided not to incorporate more complex symbol generators such as tensor products. Analysis gets small notice because the complex concepts in which it deals which are extensions of arithmetic and geometry.

In a general way, I explicitly record only systems in which the cardinal numbers of the base sets are left open. This is true of groupoids, state machines, domains and ranges of functions, order relations, and so on. But I make a few exceptions, such as indicating the integers, rational, and real numbers as order types.

My personal research has been given a larger display than it would have been by anyone else. This shows up in set-functions and the number of spaces listed. However, this may help dispel the illusion that topological spaces are excessively general. Where various systems should be placed is a matter of taste. Relations are considered by most as dominating functions. Psychologically, to me anyway, a function has direction, whereas a relation has a static geometrical flavor, in which the related elements form a pattern: Logically, relations may be represented by functions as well as functions by relations. Topological spaces and other spaces could appear under binary relations among sets as well as under set-valued set-functions. They are also representable as class-valued functions or simply as classes of sets. At this time more is known about the embedding background of set-valued set-functions than the other contexts, so I chose it.

Periodicity? I have indicated that relations include functions which include relations. Some favor resting sets on characteristic functions and making function primitive. Since binary relations include characteristic functions, then binary relations include sets, but a binary relation can be specified by its characteristic function. There are, in fact, so many inclusions and periodicities of this sort that it seemed best to put down a reasonable pattern of presenting the systems rather than getting enmeshed in what is more general than what.

Finally, this chart gives no inkling of important general principles and concepts of mathematics any more than the periodic chart of elements states other concepts, laws, and facts of chemistry. It is a skeleton without muscles, heart, or brain. But it is the first skeleton I have seen which makes organic sense.

II. SETS AND CARTESIAN PRODUCTS, FUNCTIONS, AND RELATIONS

The general plan of the master chart is to start with sets and to branch out through Cartesian product sets to functions and relations which are then subdivided into special types. One of the most important mathematical concepts of all time—that of sets—was introduced explicitly by Georg Cantor in 1897–98. A (mathematical) set, according to Cantor, is a well-defined collection of objects. While it is not possible to give a definition of "set" in this sense, it is possible to discern what Cantor meant. As used in mathematics, a set has properties which no collection of physical objects possesses. Cantor did not wish to deal with time-dependent sets of objects and, thus, he excludes collections of objects which might be referred to, for example, in using the common noun "house." He would accept, however, sets of numbers, subsets of the plane, and many others having a mathematical existence. The notion of an abstract set also originated with Cantor. If two sets can be put in a one-to-one correspondence, they are equivalent. Thus a set might be said to be an abstract set provided it can be replaced by a set of labels, for each of its elements. That is to say, the properties of the set under discussion do not involve the internal individual properties of its elements but rather the relationship among the elements. For example, from this point of view, {a,b,c} {1,2,3} and {Tom, Dick, Harry} all represent the same abstract set—one of three elements.

This concept of abstract sets is very important in mathematics since it suggests and enables emphasis on relations between elements rather than the character of the elements, which, in practical applications, might be very difficult to manage. For example, if {Tom, Dick, Harry} refers to a set of three people, considered as an abstract set, I need not concern myself with all their properties as individuals. Thus a point in a plane may be replaced in application by any one of a number of objects in practice when the plane is considered as an abstract set rather than a specified entity.

Given a certain abstract set E_1, represented by a set of actual or imagined symbols and another such set E_2, the Cartesian product set $E_1 \otimes E_2$ is usually defined by $E_1 \otimes E_2 = \{(a,b): a \epsilon E_1,\ b \epsilon E_2\}$ the set of ordered pairs of elements, the first from E_1 and the second from E_2. Actually, mathematicians do not mean this awkward way (a,b) of writing in ordered pairs but, the custom of using numerals for representing points in the Cartesian plane, for example (4,3) suggests that 43 might not be suitable. However, if we agree that ab may replace (a,b) we have the proper idea where ab is a word formed by taking letter a from alphabet E_1 and letter b from alphabet E_2.

To see more clearly what I mean by the above statement, consider three sets E_1, E_2, E_3 and then $E_1 \otimes (E_2 \otimes E_3) = \{(a,(b,c)): a \epsilon E_1,\ b \epsilon E_2,\ c \epsilon E_3\}$ and $E_1 \otimes (E_2 \otimes E_3) = \{((a,b),c): a \epsilon E_1,\ b \epsilon E_2,\ c \epsilon E_3\}$. Now these two triple products are not the same symbolically and they could be given different meanings. However, the triple product is defined by $E_1 \otimes E_2 \otimes E_3 = \{(a,b,c): a \epsilon E_1,\ b \epsilon E_2,\ c \epsilon E_3\}$ or, in linguistic form $E_1 \otimes E_2 \otimes E_3 = \{abc: a \epsilon E_1,\ b \epsilon E_2,\ c \epsilon E_3\}$. Now observe a phenomenon usually ignored. If E_1 and E_2 are abstract sets as defined above, then $E_1 \otimes E_2$ is not an abstract set since its elements have structure, i.e., (a,b) or ab. In usage only occasionally would I be allowed to treat $E_1 \otimes E_2$ as an abstract set, i.e., to replace it by another collection of symbols of equal cardinal number.

The importance of Cartesian products and Cartesian powers of sets in mathematics cannot well be overestimated. One of the most important advances in mathematics was the decimal representation of numbers, which is a linguistic way of writing Cartesian powers of $D = \{0,1,2,3,4,5,$

6,7,8,9} On the other hand, this nonabstract representation of numbers needs only consideration of using $B=\{0,1\}$ and giving binary expansions to illustrate the abstract character of numbers. Is 100 the number called one hundred or is it the number called four? It is neither, it can represent one hundred, four, or any other assigned object. It is an important use of standards, however, that they not allow many interpretations of 100 and that the decimal one prevails generally.

The introduction of higher dimensional spaces was a corollary of Descartes' use of coordinates and the necessity for dealing with many variables. Moreover, Cartesian products of two spaces are used to discuss functions and an arbitrary number of factor spaces may be used in connection with relations.

A function may be defined in any one of several ways. One way is to say that a function f assigns to each element x of a set a label or name denoted by fx. The concept of function, while not well named, is one of the best in mathematics. Any formal definition leaves out suggestions of interpretations. I now give some of my ideas concerning what functions are by the roles they play.

Initially many mathematical functions were simply formed as algorithmic expressions. For example, $fx \equiv 2x-5$ gives a way of computing the value of the function from the value of x. In many applications, the idea is that the independent variable x is one which can be measured or assumed known while the function then gave the means of converting x into a quantity which was sought. Formulas for solving triangles used in astronomy and surveying give the picture.

A function may also be thought of as a transformer or converter. It should not, I am sorry to say, be thought of as a transformation. The image here is that the function acts on an object and changes it somewhat but also some properties of the original are saved. Thus the environment of a person transforms him, electrical energy is transformed into other forms, and geometrical figures are transformed into others. The Laplace transformer changes one function into an equivalent one and this is already quite an abstract concept compared to some physical transformations.

A function is a "logic." I assume x, and fx is implied. This interpretation, while not standard, is useful. In a related way, a function is a computing machine. Put in x and out comes fx. The theory of computing is a theory of functions of certain kinds.

A function is a labeler. It does nothing to x but assign him a name. However, while this role is seemingly innocuous, the labeler may give the same name to many different objects and thus not respect the individual character of a single x. In this case functions produce all equivalence relations—two objects being equivalent if assigned the same name.

A function is a selector or a filter system. Given x, f selects or accepts fx from its range and rejects all other possibilities. This role of function, while not as passive as that of labeling, does not suggest transforming—it is more suggestive of switching devices.

A function is a connector, connecting each x to fx. Again, a function is a gauge, an appraiser or evaluater. The "size," "value," content or scope of x is fx. For example, if the inputs of f are propositions, then $fx=1$ may be read "x is true" and $fx=0$ may be read "x is false."

A function is a translator. It translates sentence x into sentence fx. A function is a mapping. It associates with geographic regions a caricature on paper.

I hope I may be excused for taking the foregone liberties, but functions are important. I next go to relations. A relation is generally regarded as a subset of a Cartesian product of a collection of sets. Thus if $R \subseteq E_1 \otimes E_2 \otimes E_3 \otimes E_4$, then $(a,b,c,d) \in R$ is read "(a,b,c,d) are R-related." This simple way of defining relations makes it easy to give a crude classification—i.e., by the number of factors in the Cartesian product of which R is a subset. Clearly a unary relation is a subset of a given set. For obvious reasons, binary relations have been the most avidly studied. They embrace graphs of functions which are then equivalent to functions, and such all-pervasive relations as equivalence, order, lattices, Boolean algebras, and so on. Relations of higher order than two are comparatively rare except in geometrical product spaces, and there, they are not usually called relations but sets because of the feeling that the coordinate system is imposed on the geometrical space and, while useful, it must not be allowed to destroy the geometrical invariants. Thus it is quite a different feeling to say that (a,b,c) are R-related than to say that (a,b,c) is a point in a subset R of E^3, i.e. the name (a,b,c) is a name of one entity, not a statement that the ordered tuple numbers (a,b,c) indicates a, b, c are related. Nevertheless, geometry will profit also by considering these sets as relations.

The theory of relations, in the general sense, is practically nonexistent at the ternary and higher levels and the principal examples are subsets of finite and infinite dimensional spaces. Professor Mesarović takes his definition of a general system to be a product space. In later sections I will discuss a few of the important binary relations.

When I considered classification of functions, I found myself on more difficult grounds than was the case in relations. The reason is that functions have domains and ranges and one cannot produce a good argument for ignoring the character of either domains or ranges. In general, I chose to concentrate on the number of independent variables, but I made one exception. The constant functions, each of which has only one value (whatever its character), seemed worthy of note. Aside from that, then, the number of independent variables became the classification device. The simple functions are functions of one variable; the finitary functions have a finite number of independent variables, and infinitary functions have an infinite number of variables.

I should call to your attention the fact that the number of variables of a function is primarily an agreement not intrinsic to what a function is. You might say that sin x is a function of one variable x. I can logically consider sin x as a function mapping an infinite dimensional space into itself. Let me assume that x is a real number represented by a binary expansion in a unique fashion. Starting from the leftmost bit, which is not zero, in the expansion of x, take it and every other one to generate the number x_1. The binary point may be left in the same place comparatively. Let x_2 be generated using every fourth bit starting with the second position from the left in x. Then the infinite sequence $x_1, x_2, \cdots, x_n, \cdots$ is uniquely determined by X. Similarly, $y=\sin x$ can be represented uniquely by a sequence $y_1, y_2, \cdots, y_n, \cdots$ determined uniquely by y. Then the function sine maps an infinite-dimensional space into an infinite-dimensional space. It is of course possible to "vary" x so that any one of the $\{x_n\}$ changes and no others. Hence the x_n are independent. Hence, if there are also numerous ways of generating an infinity of numbers from one, it is seen that sin x is a simple function only by usage, not of necessity.

In mathematical and common usage the most important functions are those with two variables because these include addition, subtraction, multiplication, division and composition, and generalizations of these. Here I should mention that functions are generators. Thus, in multiplication $f(2,3)=6$ so that 2 and 3 generate (produce) 6. While there are functions of three, four, ..., n, ... variables around, it is difficult to find a function represented which is not expressed in terms of unary or binary operators. Thus, $xy+e^z$ is a function of three variables.

The functions directly dependent on an infinity of variables arise mainly in analysis and these functions are extensions of algebraic operations, in general, using the limit concept to describe the kind of function. In this category fall infinite series, differentiation, integration. This is not to say that what the functions analysts usually discuss are functions of an infinite number of variables—it is that the processes used are infinitistic.

III. BINARY RELATIONS

Among the types of relations, as I have indicated, only those called binary (Plate 2) have a theory which is, in fact, a relation theory which is not a geometrical theory. The importance of binary relations is again hard to overstate. It is impossible to enter a discussion of mathematics without using equivalence and order assumptions. The entire notion of comparison requires at least two objects for comparison.

Now a binary relation, in general, may be a subset of the Cartesian product of two sets. These include graphs $(x, f(x))$ of functions and linguistically they include the "equivalence" between objects and their names. The usual way of discussing equivalence and order relations, however, is to assume the objects compared are in one set, say E, and hence binary relations which are subsets of $E \otimes E$ are the most-used kind.

Graph theorists claim binary relations as their field. However, this claim must be taken with a grain of salt. To claim that a given set of objects comprises the kind of thing one works with does not reveal the direction the work takes. The terminology of graph theory itself suggests a limited outlook on binary relations. However, it is fair to admit that, in principle, graph theory can be enlarged to be essentially mathematics as we know it.

Certain kinds of binary relations are indicated by name: symmetric, antisymmetric, asymmetric, reflexive, transitive, total, irreflexive. Let $R \subseteq E \otimes E$. Then R is symmetric provided (p,q) in R implies (q,p) in R. If (p,q) and (q,p) in R implies $q=p$, then R is antisymmetric (a dubious designation at best). If (p,q) in R implies (q,p) not in R, then R is asymmetric. If R contains (p,q) for every p in E, then R is reflexive, since then each p is R-related to itself. If $p \neq q$ implies at least one of (p,q) (q,p) is in R, then R is a total relation since every pair of elements of E is compared by R. Finally, if (p,p) is not in R for every p in E, then R is irreflexive. A relation R is transitive provided (p,q) (q,r) in R implies (p,r) is in R. A transitive relation is, as John Kelley (1955) points out, basically an order relation. In the chart I follow the more usual designations, which are unreasonable since a strict order relation is not an order relation.

A relation which is transitive, reflexive, and symmetric is called an equivalence relation. A (partial) order relation is an antisymmetric, reflexive transitive relation. An irreflexive transitive relation is called a strict order relation. A strict linear relation is a strict order relation which is total. Since dropping the identity relation $I = \{(p,p): p \in E\}$ from any order relation makes it strict and since incorporating the identity into any strict order relation makes it an order relation, the two kinds of order are in that sense equivalent. However, they are not equivalent when it comes to dealing with such fields as lattices where the reflexivity is necessary.

I will not engage in a discussion of order completenes except to mention that one gets the irrational numbers joined to the rational numbers by completing the set of rational numbers in the order sense. Then every set of real numbers bounded from above or below has a greatest lower bound or a least upper bound and, hence, in their usual order, the real number set is complete.

Now various types of areas are inserted under order relations to show, for one thing, how many of these have examples in linear order relations. A semilattice is simply a partial order relation in which every pair of elements $p,q \epsilon E$ has a unique upper bound in E (lower bound if you prefer). In a lattice, each pair, p,q has both a unique upper and a unique lower bound. Linear order relations are specialized to allow special places for sets of positive integers, sets of all integers, rational numbers, and real numbers as order-types.

Every binary relation is equivalent to either one of two set-valued functions. For example, let $u(q) = \{p:(p,q) \epsilon R\}$. Then the function u maps E_2 into the class of subsets of E_1 and is equivalent to R. The use of set-valued functions will be amplified later. However, binary relations among sets comes under binary relations.

I have examined in a little detail certain binary relations among sets. A. D. Wallace and R. Szymanski used certain binary relations among subsets of a space to define topologies. A binary relation R in PE, the power set of E is a separation provided (X,Y) in R and $X_1 \subseteq X$, $Y_1 \subseteq Y$ implies (X_1, Y_1) in R. Again R is exclusive provided each pair (X,Y) in R is comprised of disjoint sets. A symmetric exclusive separation R I call a Wallace separation. Let R be a Wallace separation. Then a set $A \subseteq E$ is R-connected provided $A \subseteq X \cup Y$, (X,Y) in R implies $A \subseteq X$ or $A \subseteq Y$. I have shown that this relation of connectedness is suitable and powerful in that theorems are proved which are stronger when specialized to topological spaces than those heretofore proved in topology. I have also shown that more general concepts of connectedness are necessary (Hammer, 1964).

Thus under Wallace separations and connectedness we have topological separations which are equivalent to topological spaces. Topological spaces and all the isotonic spaces listed under functions could have been placed under the more general binary relations among sets.

I mention incidence geometries under binary relations to suggest a place for them. Plane geometry as a whole is too complex to be placed here.

IV. BINARY OPERATIONS

There are functions of two variables which give values outside the domain spaces. These I ignore here and proceed to state machines and groupoids: A state machine is described as a function mapping $I \otimes E$ into E where $a \epsilon I$ is called an input and $s \epsilon E$ is called a state. Thus (a,s) generates a state $f(a,s)=s_1$, for example. The state machine is thus analogous to but more general than multiplication of vectors by scalars. I do not place groupoids in which the function maps $E \otimes E$ into E under state machines because they have a somewhat different flavor. A state machine implies a time lag in the generating of state $f(a,s)$ from a, s. Algebraists do not usually respond to such a delay concept.

Thus a groupoid is at hand whenever a function is defined on every ordered pair of elements from a set and has values in the set. One might think that such a concept is too general. On the contrary, you probably first were dealing with groupoids when you learned about substraction in the set of all integers. Substraction is not associative, it has no identity which is both left and right and hence it is not a semigroup or a loop. One's first experience with semigroups in a usual algebraic sense would involve addition or multiplication in the positive integers. However, subtraction in the positive integers or division in the same set is more general algebraically than groupoids since the operations are not defined for every pair.

Groupoids should have appreciable educative value. Given any rather small set E of elements, a child can construct his own multiplication table by filling in for each (x,y) from E an element of his choice from E. Suppose this were done for a set of 10 elements. Then since there are 100 places in the table with 10 choices for each there are 10^{100} groupoids which could result! Even discounting such obvious equivalences as those due to permuting the symbols in E, it is quite likely that the child will have written down a groupoid never recorded before! The value of this is to call attention to the impossibility of recognizing all kinds of systems and to the inherent difficulty of arithmetic where the two tables are infinite.

Of semigroups, loops, and groups, little need be said. The semigroups and groups comprise building blocks of many mathematical systems such as rings, fields, skew fields, linear vector spaces, topological groups. They also apply to virtually every mathematical system. The study of binary operations in interplay comprises a large share of abstract algebra.

V. SIMPLE FUNCTIONS

Since there seems to be a foolish superstition about functions, I take this opportunity in my chart, Plate 1, to emphasize the ordinary appear-rance of nine kinds of simple functions involving every ordered pair from {elements, sets, classes of sets}. It seems to be the case that discussion of any kind of mathematical object in a set leads to considering sets of these objects and classes of sets. For example, I may start off with a certain set of functions. Then I will want to discuss subsets of these (a family of functions), for example, all polynomials of degree at most n. Then I will want also to discuss the class of all sets of functions where each set in the class is comprised of all polynomials of a given degree. This being the case, I will find myself relating a function to a set of functions (e.g., its antiderivatives), a function to a class of sets of functions (those which form a neighborhood base), and vice versa, and so on. Not to use functional notation when the output or input or both are sets is ridiculous.

Now for the element-to-element functions there are a variety of classifications but to go into such details here is pointless. Such details should be part of a course devoted to an overview of mathematics. It is understood of course, that an element can be any kind of entity—a function, a point in a space, a set, a class of sets. The reason I go to set-valued set-functions is that I will use the fact that sets are involved and use the properties of sets to help classify functions.

As with groupoids, so it is with set-valued set-functions. They are ubiquitous since they are encountered very early both in mathematical, and common languages. The even integers can be thought of as generated by the integer 2; $f(2) = \{2n\colon n \text{ is an integer}\}$. The set of all my ancestors, the set of things referred to by a common noun, the set of factors of an integer, the antiderivatives of a function, the topological closure of a set, the boundary of a set, the zeroes of a polynomial, the set of equation systems equivalent to a given one, and so on and on. Thus set-valued functions are here and were here.

I enter upon a classification system for a limited number of set-valued set-functions which are associated with general space structures. There are here discernible two kinds of such functions. In one kind the function values must be subsets of the same space as the one containing the domain sets. The other kind does not require this restraint. The latter kind include isotonic, subadditive, additive, antitonic, superintersective, intersective, and so on. Let f map, say PE_1, the class of all subsets of E_1 into PE_2, the class of all subsets of E_2. Then the following definitions are appropriate.

1. f is isotonic provided $X_1 \subseteq X_2$ implies $fX_1 \subseteq fX_2$ i.e., f preserves inclusion. This condition is equivalent to $f(X_1 \cup X_2) \subseteq (fX_1) \cup (fX_2)$ always and to $f(X_1 \cap X_2) \subseteq (fX_1) \cap (fX_2)$ always.
2. f is subadditive provided $f(X_1 \cup X_2) \subseteq (fX_1) \cup (fX_2)$, for all X_1,

X_2 in PE_1. Note that superadditivity is equivalent to isotonicity here.

3. f is additive provided $f(X_1 \cup X_2) \equiv (fX_1) \cup (fX_2)$.
4. f is antitonic provided $X_1 \subseteq X_2$ implies $fX_2 \subseteq fX_1$ i.e., an antitonic function reverses inclusion. Equivalent definitions are $f(X_1 \cup X_2) \subseteq (fX_1) \cap (fX_2)$ always or $f(X_1 \cap X_2) \supseteq (fX_1) \cup (fX_2)$ always.
5. f is superintersective provided $f(X_1 \cap X_2) \supseteq (fX_1) \cap (fX_2)$ always. Note that subintersectivity is equivalent to isotonicity.
6. f is intersective provided $f(X_1 \cap X_2) = (fX_1) \cap (fX_2)$.

Let me next go to functions f which map PE into itself. Here the range of the function is contained in the domain. Letting F be the family of all functions mapping PE into itself, I define $f \cap g$, $f \cup g$, and $f \subseteq g$ by $(f \cap g)X \equiv (fX) \cap (gX)$, $(f \cup g)X \equiv (fX) \cup (gX)$ and the relation $f \subseteq g$ provided $fX \subseteq gX$ for all X in PE. Thus F can be treated as a class of sets. Moreover, the composition fg is defined by $(fg)X \equiv f(gX)$ which is allowable since the range (output) of g is in the domain (input) of f. Since composition of functions, when allowable is associative, F is a semigroup under composition (also under $\cup$ and $\cap$) and it has an identity element e, such that $eX \equiv X$. The complement function c is defined by $cX \equiv E \backslash X$, the set of elements in E not in X. New kinds of functions f allowable are defined below.

7. f is idempotent provided $f(fX) \equiv fX$, i.e., $f^2 = f$. Note that fX must be in the domain of f for this definition to make sense.
8. f is enlarging provided $fX \supseteq X$ always, i.e., $f \supseteq e$.
9. f is shrinking provided $fX \subseteq X$ always, i.e., $f \subseteq e$.
10. f is a limit function provided $fX \subseteq cX$ always, i.e., $f \subseteq c$.
11. f is supercomplementary provided $fX \supseteq cX$ always, i.e., $f \supseteq c$.
12. f is a primitive function provided either one of the two equivalent identities holds:

    $fX \equiv \cup \{(f \cap c)Z \colon Z \subseteq X\}$

    $fX \equiv \cap \{(f \cap c)Z \colon Z \supseteq X\}$

    A primitive function associates with a set its primary or first-order limit points.
13. f is expansive provided f is isotonic and enlarging.
14. f is contractive provided f is isotonic and shrinking.
15. f is a closure function provided f is expansive and idempotent.
16. f is an interior function provided f is contractive and idempotent.
17. f is a Kuratowski closure function provided f is an additive closure function and $f\emptyset = \emptyset$; (the null set of E is closed).
18. f is a topological interior function provided f is an intersective interior function and $fE = E$.

19. If f is any function in F then its primitive function f' is defined by: $f'X \equiv \cup \{(f \cap c)Z : Z \subseteq X\}$.
20. If f is a closure function and f' is its primitive function, then $d = f'f$ is the derived function of f.

Two functions, $f,g \epsilon F$ are dual provided $g = cfc$ (or equivalently $f = cgc$). The dual family to a subfamily G of F is defined by $cGc = \{cgc : g \epsilon G\}$. The following pairs of function types defined are dual {isotonic, isotonic}, {additive, intersective}, {subadditive, superintersective}, {antitonic, antitonic}, {idempotent, idempotent}, {limit, supercomplementary}, {enlarging, shrinking}, {primitive, primitive}, {expansive, contractive}, {closure, interior} and {Kuratowski closure, topological interior}.

Duality applies in case functions map PE_1 into PE_2 provided I use the appropriate complements, say, $g = c_0fc$ where c_0 is the complement function relative to E_2 and c the complement function relative to E_1. Duality is both a blessing and a curse. It means that every result proved involving functions on one side of a dual situation has an equivalent result on the other side. Failure to observe the formal dual equivalence in treatises leads to misunderstandings. Including it involves additional space.

VI. SPACES AND SET-VALUED FUNCTIONS

Certain spaces can be described by any one of various ways. I will first consider a general neighborhood "space" generated by an isotonic function f mapping PE_1 into PE_2. Let (f,g) be a pair of dual isotonic functions each mapping PE_1 into PE_2.

A set $X \subseteq E_1$ is called a convergent of $p \epsilon E_2$, provided $p \epsilon fX$. A set $Y \subseteq E_1$ is called a neighborhood of $p \epsilon E_2$ provided $p \epsilon gY$ (recall $g = c_0fc$ and $f = c_0gc$). The picture I suggest here is that p is an objective or goal, the neighborhoods of p are conditions which must be met to achieve p, and the convergent sets are those sets which meet all the conditions. It may readily be proved that X is a convergent of p if and only if X has a point in common with every neighborhood Y of p and, conversely Y is a neighborhood of p if and only if Y has a point in common with every convergent X of p. The effect of the isotonicity limitation here is that if X is a convergent of p and $X_1 \supseteq X$ then X_1 is a convergent of p and, similarly if Y_1 contains a neighborhood of p then Y_1 is a neighborhood of p. On suitable specializations I now get a variety of more standard spaces. In this case the neighborhoods and convergents of a point p need not be contained in the same set as p. As a matter of fact, there is no reason here why E_2 should have any other elements than p.

Now if $E_1 = E_2 = E$, then the dual isotonic pair of functions (f,g)

determines an isotonic space in E. Note that (g,f) also determines an isotonic space in which the roles of convergent and neighborhood are interchanged from that of (f,g)-space. The following special spaces are named for $E=E_1=E_2$, (f,g) a dual isotonic pair of functions.

1. If f is a primitive function, then $g=cfc$ is a primitive function and I call the space (f,g) a primitive space. Here each family $\mathscr{U}(p)$ of neighborhoods of p has a base $\mathscr{U}_0(p)$ such that $Y \in \mathscr{U}_0(p)$ implies $p \notin Y$. These spaces are hence irreflexive, i.e., p is not close to p.
2. If f is expansive, then g is contractive and the resulting space is called a Fréchet space or a reflexive isotonic space since $p \in Y$ if Y is a neighborhood of p and hence $\{p\}$ is a convergent of p.
3. If f is additive and expansive, then g is intersective and contractive and the space is called a Čech space. This kind of space was also studied early by Fréchet but it has been pursued by E. Čech and his followers. Note that Čech spaces are special Fréchet spaces.
4. If E, (f,g) is a Fréchet space and f is a closure function, then g is an interior function and the resulting space is an Appert space. Appert spaces are thus a particular kind of Fréchet space. They are very important since they include all closures such as convex hull and closures under binary and other operations.
5. A space which is both Čech and Appert misses being topological only by the possibility $f\emptyset \neq \emptyset$. I call these almost topological spaces.
6. A topological space then is a Fréchet space which is both a Čech space and an Appert space and $f\emptyset = \emptyset$.
7. If the isotonic function f is idempotent, then so is g and there exist a large number of important but comparatively unstudied idempotent spaces. For example, if u is a closure function and v an interior function (not necessarily related), then uv is an idempotent isotonic function and so is vu.
8. If E, (f,g) is a topological space and for each pair p,q of points, $p \neq q$, one has a neighborhood excluding the other, then the resulting space is called a T_0-space. That is, if $q \in f\{p\}$ then $p \notin f\{q\}$. Recall that f is now the closure of the topology. For obvious reasons, then, I call a T_0-space acyclic.
9. If E, (f,g) is a topological space and to each pair p,q of distinct points there is a neighborhood of each which excludes the other, then the space is called a T_1-space. This is equivalent to $f\{p\} = \{p\}$ for every $p \in E$, i.e., singletons are closed. Obvious T_1-spaces are subcases of T_0-spaces. Only infinite sets in a T_1-space can have limit points.
10. If E, (f,g) is a topological space and to each pair p,q, $p \neq q$, there

corresponds a disjoint pair of neighborhoods, one containing p and the other q, then the space is called a T_2 or a Hausdorff space. These spaces are prized by analysts since they relate well to compactness and to convergence.

I will not describe the other spaces such as metric and normed. These are inserted to show the degree of specialization required to reach the topological spaces isomorphic to Euclidean spaces. I end this section with an observation. It is now being demonstrated that spaces on the highest level of generality presented are useful for application. The topological spaces will not fill the roles required in convexity, in relation theory, in algebra, in logic, in numerical analysis, nor in linguistics.

VII. CONCLUSION

Now that I have finished the particular task, where can I see that improvements can be made? First of all, I think the basic concept of systemization is not a luxury but a necessity. At this time I see no better choice than sets as a starting point. From there on the number of choices proliferate. Should functions be embedded in relations directly and completely? I think not. Therefore I reach the level of classification and expansion of concepts of functions and relations. It is here I feel that the most important immediate changes can be made. The binary operations themselves are probably not elaborated enough, the binary relations might be reduced somewhat for an initial presentation. While I do not apologize for emphasizing set-valued set-functions, a subset of the details would enable one to convey the ideas. Elaboration of the Bourbaki terminology for simple functions would be helpful at an elementary level.

For educational use, of course, there also needs to be an elaboration of the "big" concepts of mathematics. I am currently working up lists of such concepts which can be used to put mathematical activities in perspective. I have dealt with three such concepts—continuity, filters, and approximation in the preceding essays. Such notions as size, exemplified inadequately in measure theory, accessibility, with subcases of distance, separation, algorithms, angles, paths, solvability, and computability are important in most areas of mathematics. They are not thus associated with special branches.

Again, the processes of combination need to be presented to show how certain mathematical systems are generated from simpler ones. Moreover, this activity may be pursued to the generation of new systems.

Finally, I come to the problems of gaps. Cursory examination of the chart will show that mathematicians have stayed close to binary operations and binary relations and that investigation of other possibilities should be made. I realize that there are several activities beyond the scope of my knowledge which may be given a place. I would be surprised, however, to learn that ternary or quaternary relations, for example, have been studied intensively enough to provide an indication of their possible merits.

Even within the binary limitation, many systems have not been studied with any adequacy. Thus binary operations of reduced scope, and transitive relations deserve intensive study. The study of set-valued functions has barely begun and here the binary aspects are scarcely in evidence. Concerning the possibility of having systems too general, I would say that if mathematicians stick to sets in the sense of Georg Cantor, excessive generality is impossible, even if much energy must be spent to establish applications. In this regard, there is missing the independent professional who can afford to go where his fancy leads without concern for his work being accepted quickly by the mathematical community.

REFERENCES

Gastl, George and P. C. Hammer. 1967. "Extended Topology: Neighborhoods and Convergents," *Proc. Coll. on Convexity,* Copenhagen, 1965, Copenhagen University Mathematical Institute pp. 104–116.

Gregory Michaud, Sister and P. C. Hammer. 1967. "Extended Topology: Transitive Relations," *Nieuw Archief voor Wiskunde* Ser. 3 **15**: 38–48.

Hammer, P. C. 1962. "Extended Topology: Additive and Sub-Additive Subfunctions of a Function," *Rend. Circ. Mat. Palermo* Ser. 2 **11**: 262–270.

Hammer, P. C. 1963. "Semispaces and the Topology of Convexity," *Proc. of Symposium in Pure Math.,* Providence. Vol. 7, pp. 305–316.

Hammer, P. C. 1964a. "Extended Topology: Connected Sets and Wallace Separations," *Portugaliae Math. 22*: 167–187.

Hammer, P. C. 1964b. "The Role and Nature of Mathematics," *Mathematics Teacher* **52**: 514–521.

Hammer, P. C. 1966. "Extended Topology: Set-valued Set-Functions," *Nieuw Archief voor Wiskunde* Ser. 3 **10**: 55–77.

Hammer, P. C. 1967a. "Isotonic Spaces in Convexity," *Proc. Coll. on Convexity,* Copenhagen, 1965. pp. 132–141.

Hammer, P. C. 1967b. "Extended Topology: Continuity II," *Proc. of Topology Symposium,* German Academy of Sciences, Berlin.

Hammer, P. C. 1969. "Filters in General," this volume, pp. 107–120.

Kelley, John L. 1955. *General Topology.* Van Nostrand, New York.

Wallace, A. D. 1941. "Separation Spaces," *Annals of Math.* Ser. 2 **42**: 687–697.

8

PROGRAMMING SYSTEMS

ALAN J. PERLIS

I. INTRODUCTION

In every activity the management of complexity is arranged by use of systems. In the context in which "system" is to be used in this paper, i.e., "programming systems," there is little virtue in giving it a definition that will have a classificatory or set-membership function. Programming systems are defined as constructions operating on, or intended for operation on, computers. Their function is to control the execution of other programs, and indeed, wide classes of programs. Furthermore, rarely do any of these controlled programs use more than a fraction of the capabilities of these controlling system programs.

As is common to all systems, the design of a programming system makes some complex functions available with ease and others, seemingly little different, available only in a most grotesque way. Such faults of design are invevitable in even the best systems and are a simple consequence of the inevitable disparity between the connectivity of human thought processes and those of a fixed system. Put another way, every system displays through usage a set of unanticipated bottlenecks. To avoid these bottlenecks, users of systems learn to adjust to the system. Every programming system imposes an "etiquette" on users, as every system which is built is itself a recognition of an "etiquette."

From the short history of programming usage, certain standard kinds of programming systems have come to be defined. They are: assembly, interpreter, translator or compiler, file processing, macro, library, operating, input/output, conversational, symbol-manipulating, formula-manipulating, simulation, and data structure systems, etc. Not all of these are independent in function and definition. Yet they all serve slightly different purposes, and in each category there are large numbers of tasks which more naturally belong there than in any of the other categories.

Nevertheless, they all have a certain fundamental view in common: each such system has a language for communication with it and within it and each is designed and modeled as a machine or, if you will, a computer. It is from the three aspects of function, language, and "background machine" that these systems can best be understood.

Everyone reading these notes is familiar with a programming language and even many of the functions of programming systems, but probably most are not aware of these systems as machines, which is the only way they can be understood. However, it is well known that they can be used without being well understood and this is an important property of systems and machines.

Abstract machines, while different, are created of similar abstractions: data storage, instruction sets, and control functions, i.e., they are like the computer on which they operate.

It will not be possible in this short space to treat, even briefly, all of these types of systems. Those systems which are treated will only be examined from a limited point of view which is chosen to emphasize the most important aspects of the system.

II. ASSEMBLY SYSTEMS

Basic to every computer system and underlying every programming system is the assembly system. Generally one thinks of its programming language as a "machinelike" language corresponding to the machine on which it runs, though it will be apparent that the processes of assembly are applicable to any programming system.

The basic function of an assembly system is to organize a systematic substitution of code text and addresses for symbols. To an important degree the purpose of these substitutions is to "free" a program from irrelevant bindings, e.g., location in memory of code and data, names of locations, etc.

To visualize the system processing, we proceed in stages. First the program being assembled is considered to be a file, f, all of whose records γ have the same format. We illustrate by specializing γ to a triplet λ,ω,α (which we may correspond with location, operation, address, respectively), each of whose domain consists of symbols and constants. By signal or syntax, the set of constants is differentiated from the set of symbols. Symbols get translated into addresses. Among the constants are usually to be found integers, bit strings, character strings, *and* addresses.

The process of assembly terminates when the program is "bound," i.e., when all symbols are replaced by constants in such a way that certain relations holding among symbols are preserved by their respective constants.

Symbolically

$$f_i\colon \lambda,\omega,\alpha\rightarrow\bar{\lambda},\bar{\omega},\alpha$$

where $\bar{\lambda},\bar{\omega},\bar{\alpha}$ are constants and f_i is the ith record of the file f.

What are the relations? Let θ be an arbitrary symbol, and $>$ mean "immediately follow in file." The following are standard, though important exceptions may be programmed.

$$\text{(i) } \theta_1=\theta_2\rightarrow\bar{\theta}_1=\bar{\theta}_2$$

$$\text{(ii) } \lambda_2>\lambda_1\rightarrow\bar{\lambda}_2=\bar{\lambda}_1+c$$

$$\text{(iii) } \theta_1\neq\theta_2\rightarrow\bar{\theta}_1\neq\bar{\theta}_2$$

c is a constant we may take to be 1. Condition (ii) maintains the order of the file in the mapped version. The mapping of ω is generally fixed once and for all in the system:

$$\bar{\omega}=t(\omega)$$

because the computer has a fixed set of operations. Since they are the most restrictive transformations, λ and ω are usually mapped first.

$$f\colon \lambda,\omega,\alpha\rightarrow\bar{\lambda},\bar{\omega},\alpha\ :f'$$

followed by:

$$f'\colon \bar{\lambda},\bar{\omega},\alpha\rightarrow\bar{\lambda},\bar{\omega},\bar{\alpha}\ :f''$$

Most assemblers are of this "2-pass variety.

Among the α^s, there are the following:

(i) symbols which are λ's

(ii) symbols which are not

(iii) constants

On the first pass type (i) symbols are assigned.

On the second pass type (ii) symbols are assigned.

On the first pass type (iii) constants are pooled, i.e., a λ is created for each distinct constant, and is substituted as an α for each occurrence of the constant. For each constant a record $\lambda\mu$ is appended to the file where μ is constant. On the second pass all μ's are translated.

Generally the mapping of α is arranged so that, if m=number of different symbols, $\bar{j}\leq\bar{\alpha}\leq\bar{j}+m-1$ where j is some address constant. Since so much of machine language programming deals with vectors, an α may be an expression of the form $\theta\cdot\bar{q}$ and maps into $\bar{\theta}+\bar{q}$. It is normal to omit a λ. The next occurring λ,λ^1, maps as $\bar{\lambda}^1=\bar{\lambda}_0+n+1$ where n is the number of consecutive records of blank λ field occurring between λ_0 and λ^1. Among the records there may be those which are not to be transformed on that particular pass, though they may be on a subsequent one. While there are many ways to impose such a control on records, we will use a select set of prefixes $\{\zeta_k\}$ to indicate that a record is invisible to a pass:

$$\zeta_i\rightarrow\zeta_j$$

The occurrence of a ζ_i causes the remainder of the records to be untouched and changes ζ_i to ζ_j. A subset of the $\{\zeta_i\}$ is attached to a pass.

This is the most elementary kind of assembler and a recognition of avenues of increasing generality leads to a view of an assembly system. We can see that we can characterize the operation of this elementary assembler as specifying an operation on three parameters: f, j, f''. Generalizing the assembler can best be thought of in terms of its embedding in a larger system which is a more powerful processor. In this larger system, the parts of an assembler are themselves subject to manipulation and a more elaborate control language and sequencing rule arises for producing assemblies. Consider some of the avenues of generalization:

The formats γ and γ'' of f and f''.

The mapping function t.

The general mapping of λ and α.

However, we do still insist on the strict file sequencing control as already outlined: During a pass if $\gamma' > \gamma$ then γ is not processed after γ'.

The most important generalizations deal with the combining of files and the resulting sensitivity of control required of the γ and α mappings. An important file is a library file G. G is, of course a file of files g. The role of a library file is to provide standard "codes" which can be "linked with" an input file, f. Matters are simplified if the format and mappings of g are the same as for f, though it is clearly not necessary. As soon as G is created certain super system functions come into being. These include: editing G (addition and deletion of and within files g), and accessing g.

The file f must permit a code which calls for an access of g and even an editing of such a g, and clearly files from g possess records with the same property. Let us add the particular operator lib to ω to specify a link to a library file, e.g., λ, lib,q links to a library file q. Furthermore let us add to the symbols λ, lib*q to indicate a place in f into which $q \varepsilon g$ is to be substituted. Let s be the length of q; then $\lambda' >$ lib*$q \rightarrow \bar{\lambda}' = \bar{\lambda} + s$. Furthermore, for each distinct q for which there is an occurrence of a lib,q not balanced by a $\lambda =$ lib*q, a distinct symbol λ is created of the form lib*q.

It should be observed that the use of a λ lib*q implies that the substitution of q is to be done at this point in the file.

The occurrence of a $\lambda =$ lib*q may appear more than once in the file f. At each such occurrence the file q is to be substituted into f. The occurrence of lib,q which may appear more than once in f is meant to lead to one occurrence adjoined to f of q from all these sources. Because of their possible reoccurrence in f and in files h appended to f, all files named through lib*h, i.e., a subset of G, must be considered as available in independent files.

Upon completion of pass 1 this set of files is constructed but, because of the levels of q, only a single scan of G is required. Thus at the conclusion of pass 1 f has been mapped into $f'=f$ followed by $\{\lambda=\text{lib}*h\}$ and a set of files $\{h^1, h^2, \cdots, h_k\}$. A pass through f' is now carried out which substitutes copies of the files h_i into f'. Each such substitution has an effect on λ occurring later in the program. Upon completion of this pass all λ are bound. A final pass binds the $\{\alpha\}$.

Since a q may itself contain symbols of the form lib$*q'$ or lib,q' to maintain file sequencing of G we impose an ordering of G as follows:

> All files of G which contain no $\lambda'=\text{lib}*q$ or lib,q (other than involving their own q) are of level 0. Let the maximum level of all outside library references in a file h of G be m, then the level of h is $m+1$. Let the file G be ordered in decreasing level numbers.

Then, no file h required by a record in f' will require a back spacing through G. Of course this collation is of little value unless an appropriate communication can be arranged between code in f and the files h_1, $h_2, \cdots, h_k$ added to it.

A major concern in arranging such communication is symbol interference: If $\theta\varepsilon f$ and $\theta'\varepsilon h_j$, what can be their relation if $\theta=\theta'$ or $\theta\neq\theta'$?

Generally one proceeds as follows: Prefix to every θ occurring in h_k the symbol h_k, i.e.,

$$\theta\varepsilon h_k \rightarrow h_k\cdot\theta$$

and now set the integer j for each h as follows: Let k be the index of h_r, i.e., h_r is the kth library file added to f. Let $|p|$ denote the length of the file p. Then we compute:

$$j_1=j+|p|$$
$$j_{k+1}=j_k+|h_r|$$

The communication is now limited to execution time transfers and data references.

This process is generally called qualifying the symbol θ by h_k and will imply that $\theta\varepsilon h_k$ and $\theta\varepsilon h_l$ and $l\neq k$ guarantees

$$\bar{\theta}(\varepsilon h_k)\neq\bar{\theta}(\varepsilon h_l).$$

However, we must also provide for symbol communication: For $\theta\varepsilon h_k$ and $\Psi\varepsilon h_l$, $\bar{\theta}=\bar{\Psi}$

These equalities are usually called equivalences but are poorly called that because the relation is not symmetric. Let us use the notation $h_k\cdot\theta>h_l\cdot\Psi$ to mean the assignment to $h_k\cdot\theta$ fixes that of $h_l\cdot\Psi$ regardless of the levels of h_k and $h_l\cdot h_k\cdot\theta>h_l\cdot\Psi$ implies that

$$\overline{h_k\cdot\theta=h_l\cdot\Psi}$$

and that $h_k\cdot\theta$ is transformed before $h_l\cdot\Psi$. We say that $h_k\cdot\theta$ dominates $h_l\cdot\Psi$ and that θ is the dominating symbol. Furthermore, the dominated symbol is either in the same file as its dominant or in a file of lower level than its dominant. These relations may occur anywhere in f and

$\{h\}$ except that any relation having a symbol which occurs as a λ must occur in f' before its first occurrence as a λ. The effect of such a relation, for example, is to replace all references to $h_l \cdot \Psi$ by references to $h_k \cdot \theta$ If both symbols occur as λ's, only the one of the higher level will survive past pass 2. If they are the same level, only the left symbol will survive.

$>$assigns an address to the right symbol. Can we give an interpretation to more than one occurrence of a dominated Ψ? Yes. At the end of pass 1.5 all f and h_i called are in place: $f = h_0 < h_1 < h_2 \cdots < h_t$ with m and $n \exists n > m \rightarrow \text{level}\,(h_n) \leq \text{level}\,(h_m)$. Let us have created a file of dominances sorted by the level on the dominating symbols. Then any occurrence of $\theta > \Psi$ will find θ already assigned and so assign Ψ. Repeats of $> \Psi$ will merely reassign Ψ. Reassignments of θ will not affect Ψ unless accompanied by a further occurrence of $\theta > \Psi$.

This interfile communication is nonparametric and is certainly fixed by the end of pass 2. We need a different model for procedural parameter communication, i.e., one having more variability at the execution phase of a program. We will take that up later. All of the mappings are but slightly varied for symbols of the form $h_r \cdot \theta \cdot \bar{k}$.

The next system expansion deals with the insertion of variables into the files which are operands during file processing, and a corresponding inclusion of a set of pertinent operators and relations. Control is still dominated by file sequencing, and branching commands will have the effect of file skipping. Such records will not occur in f', and are tagged to indicate that they are to be executed (and assembled) when encountered in the pass through f.

The major function of these commands is to control the content and size or elasticity of the files f' and f''. The variables take on as values: Booleans, integers, and addresses. Generally speaking, the variables take on addresses or integers (for counting) or true or false as values.

Every symbol λ, ω, α can be considered a variable, and a value is assigned by the mapping function. Consider now a symbol of the form

$$\lambda - \text{lib}^{*}h \qquad (\text{or}^{*}h, \text{ for short})$$

The assembler has assumed h is a constant. Suppose we allow σh, i.e., h is a variable. Then an assignment:

$$\theta > h$$

must occur "before" the file can be selected, i.e., as file θ. Again, the dominance must obey the rule of levels. It is of course possible to insist that all references to h occur before ${}^{*}\sigma h$ is encountered and thereby improve efficiency but in an initial system design this is unduly restrictive and unwise. Thus we see that it is now possible to have variable library files, which of course may themselves contain variables. This control sequence produces $\lambda\omega\alpha$ sequences which are to be assembled relative to the current value of λ.

Let us add the prefix ξ to a library call, e.g., as lib*ξh or lib*$\xi\sigma h$. It is understood that such a library file is to be assembled, relative to the current value of λ, the current file (f') being held fixed, and then executed to completion before resuming the assembly of the current file (f'). Such files communicate with the current file by means of quoted instructions which, instead of being executed, are substituted at the current value of λ (which is of course then incremented), and, of course, by means of dominances.

However, note that within a file called by ξ, other files within it called by ξ place their quoted instructions always relative to f' and not relative to any ξh file. Within a ξh file a conditional instruction $\tilde{\omega}$ may appear which is of the form $\tilde{\omega}\alpha_1\alpha_2$ and tests whether $\alpha_1 > \alpha_2$. An $\tilde{\omega}$ instruction occurring in a file f is ignored. Such an instruction permits control over the insertions which will be made into f.

The act of insertion of text is one which causes the inserted text to conform to that of the parent file. This conforming is arranged by a specific set of dominances. It is customary to have such an h in the library to be headed with a sequence of its symbols which are to be dominated and the lib*ξh to be immediately preceded by a list of symbols which will correspondingly dominate them. These sequences are called the formal parameter and actual parameter sequences, respectively. A notation is lib*$\xi h(\alpha_1,\alpha_2, \cdots,\alpha_k)$ for the call and $h(\beta_1,\beta_2, \cdots,\beta_k)$ as a header in the file h. These substituted files ξh are commonly called macros.

We now turn to the question of entering files into G. Within the file f may appear an "anti-quote" ω,τ, bracketing a set of consecutive records, i.e., a record sequence of the form:

$$\lambda\tau h(\cdots \lambda\tau)$$

Such a sequence is to be deleted from f and entered into G as the file h. However, such a file is a "constant file" and has not been subject to assembling. Let us agree to put a final code on our file f which specifies the nature of pass 2. One form of pass 2 is to complete the assembly by binding $\{\alpha\}$ according to the mappings already described. But another form of pass 2 is possible:

Replace every $\bar{\lambda}$ by $\bar{\lambda} \cdot j$.

Replace every α by its dominant.

Form a sequence j, ν_1, ν_2, $\cdots$,ν_k where $\{\nu_i\}$ are all symbols which dominate but are not themselves dominated.

Affix the name h as given in the final instruction of f.

Insert h into G. It will be a library file which has been edited by the assembly process and has a level 0.

More control over quoting and antiquoting is required so that λ lib*h are not executed. The h_i controls can be used for this purpose. The file can be entered into the library anywhere preceding all of h_1, h_2, $\cdots$, h_k, i.e., the file of the library need only be partially ordered, i.e., in the first available file space in the library file. Note that j has been absorbed into the file as a symbol. Put another way, the relocation base j is only assigned at pass 2; it is a pass 2 variable. The setting of all "external" variables is a pass 2 action, e.g., the base location j, the particular resources (tape, printer, drum) chosen. Among the possible choices is that made by the system, particularly useful if the choice is irrelevant to the programmer. The accomplishment of pass 2 is a separate activity of the assembly system and is always a part of an execution, whereas pass 1 is a separate activity concerned with assembly. This is the classical view, and a better one would be: assembly, loading, execution.

We do not choose to introduce more complex syntax into ξ sequences, as is customary with macros, until it is introduced for all $\lambda\omega\alpha$ sequences. There are some points of difference between the system described and those customarily found in practice. They have primarily to do with syntax. We shall take that up shortly.

In summary of this section let us recount the major features:

The system is "doubly productive": it produces translated sequences and it produces library sequences of both the assembled and unassembled kind.

The system (or process) is "open," to a degree, i.e., the sequences on which it operates can themselves control, to a degree, the manner in which the system processes them. This is, of course, the role of variables in the sequence.

The system is "contained," i.e., there is another system containing it in which the translated sequences are processed, i.e., the assembler and G.

The system is a machine and it can be effectively used only when its controls and variables are understood.

III. COMPILER SYSTEMS

Once an assembly system has been designed, extensions of one sort or another can be designed and added. A most frequent one is syntax extension. This proceeds in two phases: fixed and variable. Furthermore, we find in examining the development of these later systems that a curious inversion has occurred: Solution of the complexities arising from expansion of syntax tends to dominate the attention of system designers over the complexities arising from symbol assignment even though the two complexities are not antithetical. This is not unnatural since the sequence writer interfaces with the syntax—supposedly adapted to his needs—and generally eschews complex symbol assignments and assembly control. "High-level" language systems which provide the programmer with adequate control over assemblies are, however, beginning to emerge as conversational programming systems.

While it is not the intent in these lectures to describe "algebraic" and "symbol manipulation" languages, it will be appropriate to say a few words about the major issues.

A. Expressions

As epitomized by arithmetic expressions, these sequences of operators and operands "express" control arising from composition in a most concise way. In a language dominated by tasks of an arithmetic nature, the use of arithmetic expressions is essential. Infix notation designed for binary and unary operators provides a further condensing of notation.

Control is always defined by imputing a direction of scan of the expression sequence (left to right or vice versa) and, on that, often a heirarchy or precedence which can, in most common cases, eliminate explicit control characters like the parenthesis. Expressions yield a value, and anywhere in the language that such values may appear, expressions may appear. Conversely, every form which takes on a value is an expression.

B. Control

Certain standard types of control can be attractively expressed in elegant syntax.

Replication. A certain set of actions is to be repeated "under control." The issues are: delineation of the set of actions, the variation of the control during each cycle, and the setting of the terminal condition. Usually the set of actions being controlled is defined by lexicographic connectivity but other controls are possible, e.g., all actions for which a state variable has a certain value. Generally, lexicographic connectivity is the natural way of arranging the set.

Branching. This fundamental action can be nicely stated in expression form as a selector of one of a pair of subexpressions. The selector, instead of mapping onto a range of two values (true (0) or false (1)) can be arranged to map onto the integers $0,1, \cdots ,k$ and select one of $k+1$ alternates. The introduction of Boolean variables (true and false domain) can only be tolerated on the basis of storage efficiency. Assigning an integer value to a relation is so much more flexible—and does cut away one variable class.

The Subroutine. The development of this old programming concept in languages of rich syntax has been exceptional. The ALGOL procedure opened a rich vein of programmer control, particularly in the use of recursion. A word should be said here about procedures. These are connected sequences of code which are named and parametrized. They are like the macro expansions we have described except that they are executed after assembly.

It must be realized that procedures are shared processes. As such, they have both a private invariance and a public control. Generally the nature of both is fixed in the procedure design. The public control is expressed in the form of the so-called assignment of actual (public) parameters to the formal parameters. The formal parameters are templates and the greater the range of their substituends the more flexible the procedure. However, most syntax-oriented systems tend to require at least an invariance of syntax in the assignment, e.g., a formal which plays the role of an expression must always be replaced by an actual which is an expression. Very few procedure systems permit the number of parameters to change from call to call.

The control of communication permitted by the actual parameters has become more varied in recent years.

The actual parameter is a constant—not in form necessarily, but the public communication is cut off after an initialization.

The actual parameter is itself a procedure and the public communication is continuous throughout the execution of the procedure.

In either case the parameters may take on values from the domain of locations (links or references) or problem variables.

The foregoing discussion has emphasized how the preoccupation has altered. We might say that the first system was involved in the preparation of symbolic sequences and this second syntax-oriented system with the execution of symbolic sequences. Historically the first is called an assembly system and the second a compiler system. Systems in which both activities are important and occur at all times are generally called interpretive systems. These latter systems, out of favor until recently, have become important with the advent of time-sharing and conversational computing.

IV. CONVERSATIONAL SYSTEMS

Parallelism in computer systems is best managed for user programs by a program variously labeled: "supervisor," "monitor," "executive," etc. The reasons are both sociological and technical and arise from the variable load on the machine resources created by any single program and the difficulty of any single program in taking full advantage of parallelism. Thus the resources are most efficiently shared among several programs, each of which is in a running state. This further supports the need for an executive to guarantee that each program executes as though the other programs were not present, i.e., they are mutually unobtrusive. Time-sharing is a view of this multiprogramming environment which accents large numbers of programs being simultaneously in a running state and for which the response of the system to individual computation requests is rapid.

In summation, the purpose of time-sharing is twofold: to increase the efficiency of the computer system and, while attaining this increase, to permit efficient computer communication between a programmer and his programs. This communication we may call conversation. Prevailing programming languages like FORTRAN, PL/I, ALGOL, COBOL, etc. are poorly designed for such interactive (i.e., conversational) programming. However, languages like JOSS[1], AZL[2], and LC[2], to be described, are much more suited to this task.

The properties of a conversational language system may be described as follows:

1. Rapid response to trivial requests.
   A conversational system should include a "desk calculator" facility which allows simple calculations and editing functions to be done with no noticeable delays.
2. Pacing of the conversation by either the programmer or his program, arbitrarily nested.
   A conversation implies not only the exchange of information but control over this exchange exercised by either party. This control may be different at different times, and, since programs are being dealt with, the whole process is recursive, being managed differently at different depths of the program being executed.
3. Reminder-type record keeping.
   Since text may be added to at any time, incompletely specified programs need not be erroneous. Thus a list of unspecified program steps entered as a program is a reminder of precisely those steps which must be done to complete the program, and they need not be specified until they are to be first executed. But these steps are a flow chart, and the execution of the chart paces the act of providing the detailed specification.
4. Tentative or explorative execution of tasks.
   Procedures are successful (i.e., correct and useful) parametrizations of blocks of program text. A role of conversational programming is the finding of procedures. Hence blocks of text should be conveniently executable under different environments and parametrizations. In particular, it should be possible to execute a block of text in such a mode that all changes to nonlocal data and

text arising during execution of the block can be summarily revoked immediately upon exit from the block.

5. Flexible composition rules.
6. Ability to retain programs and data in a variety of states for an indefinite amount of time.
7. Condensed notation.
   Brevity improves conversation. In most cases a conversation is private and the exchange is of transitory value. Hence the notations for conversational languages can be more brief and possibly less transparent to casual reading than the "prose" text of ALGOL or PL/I. Furthermore, the balance holding between the complexity of operators and their coded representations changes for conversational languages.
8. Multiple notations.
9. Natural, possibly even mechanical, transference of programs to a nonconversational language such as FORTRAN or ALGOL.
   Programs are often created which are to enter the public domain, such as algorithms. Hence there should be a natural and mechanical means of translation from the records of a conversation to the "prose" of ALGOL or PL/I.

A conversational language is a system organized to supply the power of the language to the user. In what follows system and language are considered as one, though they are in fact separable in many ways. The key to conversational systems is the reliance on an interpreter rather than a translator to process the text of the program. While the differences between interpretation and translation can be made as fuzzy as one wishes, qualitative differences of consequence arising from interpretative control can appear in the language:

1. The program text, always being under direct execution, is understood to be "immediately" subject to change and hence the system statements known as editing statements are part of the language. Put another way, the text is a data type of the language.
2. The order of text entry, i.e., the order of its creation, need not be as rigid as required by most translation systems; indeed, the order is irrelevant.
3. The scope of identifiers and the definition of a procedure can be much more dynamic.
4. The states of program modification and execution can be arbitrarily and dynamically laced.
5. Program variables may naturally take on values of different types during stages of execution.
6. The system can arrange the execution of program parts in protected environments.

A. The Conversational Language LC²

LC² is a language for conversational computing which is being implemented within the TSS monitor system on an IBM 360/67 computer at Carnegie-Mellon University. In its design, LC² is basically an amalgamation of the basic elements and statements of ALGOL[3] and the input-output, control, editing, and filing statements of JOSS, but the control structures have been extensively modified to give a user a very general kind of dynamic control. Space does not permit a complete description of LC² here, but its most important features will be described. Statements in LC² may be used in two different ways, either immediate or delayed. Delayed statements are translated and saved inside the computer, from where they may be recalled and executed under programmer control. A set of delayed basic statements which are typed in together and are to be treated as a single control unit is called a step. It is distinguished by the presence of a preceding decimal step number which indicates its relationship to other steps. Steps are ordered according to their numbers, and they may be freely inserted, modified, or deleted while conversing. Steps can be typed in any order, and any newly entered step will replace a previously saved step with the same number. For execution purposes, steps are grouped into parts, with a part being the ordered set of all steps whose numbers have the same integer position.

An immediate statement, which does not have a step number or a label, is translated and executed when typed and is then discarded. Immediate statements are used to perform one-time or "desk calculator" calculations, to control the execution of the steps of a program, and to perform various editing and debugging functions. Any basic statement form may be used in either the immediate or delayed mode.

A part in LC² may be called by a reference to its name (e.g., PART 6). If the call is made in an immediate step by the user, then the program gains control; this is the primary method of executing a set of delayed steps. In effect, the immediate step gains control until it has finished execution. A part may be called in a delayed step such as

3.2 PART 2;

in which case the ordered steps of part 2 would be executed as if they were the code of an ALGOL procedure without parameters, and control would return to the step following 3.2 when part 2 terminates.

With respect to the scope of names, an LC² part is similar to an ALGOL block in that it is a program segment within which a declaration holds. However, there are fundamental differences due to the desire to make LC² dynamic. First of all, a declaration is treated as a statement whose effect is not realized until it is executed. Thus a declaration may appear anywhere in a part and it may be executed conditionally. An identifier may even be declared more than once in a part, but its meaning will always be that associated with its most recent declaration. Second, a part can only be executed by being called as a closed routine, i.e., control cannot "run into" a part. Third, a nonlocal identifier (one which has not been declared in a part) will represent the same entity inside the part as it did in the context which existed immediately before the part was called. Thus a nonlocal identifier in a part can represent different entities, depending on the call of the part, and, if declarations intevene, it can even represent different entities on two successive executions of the same call.

The semantics of such dynamic scope of names is best given by an explanation of a possible implementation. Consider each variable in LC² as a stack of incarnation values, each composed of a level number corresponding to the dynamic nesting of parts due to calls and a value for that incarnation. Each entry in a table of variable names (the symbol table) points to the top entry of the incarnation-value stack for that variable.

For each variable name, there is always at least one entry in its stack at level 0. Level 0 is called the global level and is the initial state when the user starts with the system. As soon as the translator encounters an identifier (it need not be declared), a global declaration is established for it and a level 0 entry placed in its incarnation-value stack. Whenever a part is called, the current level number of the execution increases by 1, and when control leaves a part, e.g., by a RETURN statement, the current level number is decreased by 1. Thus, if a variable X is declared in some part at execution level *K*, a new entry is placed on *X*'s stack. Normal accessing of a variable involves only the value on the top of that variable's stack, so whenever *X* is accessed at level *K*, the currently declared value of *X* will be used. Upon exiting from the part in which *X* was declared, the current level number is decremented by 1 and all incarnation-value entries whose level numbers are greater than the new current level number are deleted from their respective stacks. Thus the previous incarnation of *X* would be the one used after a part in which it was declared had returned control. If *X* has been declared at level *K*, and another declaration of *X* is executed at the same level, the new declaration simply replaces the old on the top of the stack and the old level *K* declaration disappears. The following sample portion of an LC² program session shows the effects of such dynamic nesting.

```
REAL A←5; COMMENT A IS DECLARED REAL AND ASSIGNED THE VALUE 5;
STRING B←'GLOBAL B';
COMMENT PART 7 WILL ONLY HAVE ONE STEP;
7.2 TYPE A, B, C; RETURN;
COMMENT PART 2 WILL ASSIGN VALUES TO A AND C AND THEN CALL
```

PART 7;
2.1 STRING A←'PART 2:A';
2.3 C←2; COMMENT C WILL BE IMPLICITLY REAL AND GLOBAL:
2.5 PART 7; RETURN;
TYPE A, B; COMMENT C STILL HAS NO VALUE BECAUSE STEP 2.3 HAS NOT BEEN EXECUTED;

A =5
B =GLOBAL B

PART 2; COMMENT PART 7 WILL GET ITS VALUES WITHIN THE CONTEXT OF PART 2, WHICH CALLED IT;

A =PART 2:A
B =GLOBAL B
C =2

PART 7; COMMENT NOW PART 7 WILL FIND ONLY GLOBAL VALUES. THE A WHICH WAS DECLARED IN PART 2 WAS "POPPED" AND LOST WHEN THE PART WAS CLOSED;

A=5
B=GLOBAL B
C=2

As was mentioned, LC² derives a great deal of its power from being interpretive. Interpretation can be exploited in the handling of functions (or procedures or parts). Suppose one has created a part, say, PART 4. Its creation or definition is a static act, i.e., one which involves no evaluation. This is much like entering a library file into the library, as described in Section I.

However, it is important to be able to specialize such texts dynamically in the definition of other functions. The specialization of a PART, e.g., PART 4, may be considered as a dynamic act

$$\text{LET } P \leftarrow M \cdot \text{PART } 4$$

where P is the new text produced, M is an editing control and PART 4 is the text being altered. When this is executed is P-define time. When P is executed we will call P-execute time. At any one time only one P can be executed and only one P can be defined, though

$$\text{LET } P \leftarrow M \cdot P$$

is possible.

The symbols in PART 4 not only have (current) values, they may have many incarnations. Which are to appear in P? It is the role of M to specialize PART 4 to P, and the specialization is the execution of M, actually itself a part.

In keeping with the generality of section I M should be LC² code, but as (ruefully) indicated there the increase in syntax of LC² has caused a decrease in the generality of M. M has the following instructions which we specify by example:

(i) $\nabla(U,V)$ The symbols U,V are isolated as formal parameters of P, i.e., it can be called as $P(\xi_1, \xi_2)$ where ξ_1 and ξ_2 are then actual parameters of P.

(ii) *c*w The symbol w is replaced by its current values

(iii) *i*w The symbol w is anchored to its current incarnation.

(iv) A symbol may be assigned a value

w←*c*k+*i*x

(v) Any editing statement of LC² (usually intended to apply to PART 4) may be employed. The modifications will appear in P.

(vi) Any of PART 4 can be specifically executed—in the current context—by the use of the execute statement.

PART 4 may itself contain ∇ lists. Execution of P must supply the values of all such variables and an ordering convention is defined so that an unambiguous pairing is possible.

As has been indicated, a part is the body of a procedure without parameters. In LC² parameters can be specified to such a procedure body, either by a LET statement or a call. A LET statement declares an identifier as the name of a part and may associate some formal parameters with that name. An example is

16.3 LET $F:= \quad \nabla(X,Y,Z)$ PART 4;

A subsequent call might be

7.3 $F(A-B, C[I], 15)$;

The effect of this call would be to first push down the current incarnation values of X, Y, and Z and then to initialize their new incarnations to the actual parameters $A-B$, $C[I]$, and 15 respectively. In PART 4, references to X,Y, and Z, which should all be nonlocal, will be to these actual parameters. Parameter assignments will normally be by name, but they can be explicitly specified as "by name" or "by value" in the call. Thus,

7.4 $F(=A-B, C[I], \leftarrow 15)$;

has the effect of making Z a value parameter and X and Y name parameters. Note that in different calls of the same procedure, the kinds of assignment can be different; for example,

7.7 $F(\leftarrow P+Q, \leftarrow 10, =X)$;

To change the parameters to a part from call to call, the parameter assignments can be made in the call. This is done by giving a list of explicit assignments to new incarnations of some of the identifiers used in the part being called. An example is

PART 4 $\{X=A-B;\ Y=C[I];\ Z\leftarrow 15\}$;

which would have the same effect as the combination of steps 16.3 and 7.4 above. These two kinds of call may also be combined, e.g.,

7.83 $F(=A-B, C[I], \leftarrow 15)\ \{U=V*W[J]\}$;

Normally, a part will terminate by executing a RETURN statement. Examples of RETURN statements are

11.3 RETURN;

11.4 RETURN $(A+B)$;

A RETURN statement will cause control to return to the calling statement and will undo any and all effects of incarnations made in the returning part or its call. If an expression is included in the RETURN statement, its value is returned as the value of the part, which should then have been called as a function.

During the construction of a procedure in a program session, it is often desirable to be able to execute an untested part in a protected environment, i.e., one in which nonlocal variables are not changed during the part's execution. This is done by modifying the call to SET the identifiers which are to be protected. An example is

146.53 $G(X-Y, Z)\{U=V[K];$ SET $P,Q\}$;

The effect of the SET is to push down the current incarnations of P and Q and then copy the pushed-down incarnation values as the new ones. Within the part, execution will proceed exactly the same as if a SET had not been present, but if any changes are made in P or Q, only their local copies will be changed. On leaving the part, these local values will be lost and the original values for P and Q restored. If the user finds that the part works correctly, he can undo the action of the SET by including a RESET list in a RETURN statement. As an example, the statement

4.932 RETURN $(CDE\text{-}5.6)\{$RESET $P,Q\}$;

would, as usual, delete all local incarnations, but would do so only after copying the local values for P and Q into the entries below them in their respective stacks. Thus any variable which is both SET and RESET in the execution of a part will behave operationally as if neither SET nor RESET had been used.

An example containing some simple declarations and calls is given below.

```
6.1 COMMENT THIS IS EQUIVALENT TO THE ALGOL 60 INNERPRODUCT
PROCEDURE BODY:
6.2 REAL S←0;
6.3 FOR P←1 STEP 1 UNTIL K DO S←S+A*B;
6.4 Y←S;
6.5 RETURN;
COMMENT WE CAN GIVE THIS PART A NAME AND SPECIFY SOME FORMAL
PARAMETERS;
LET INNERPROD:=∇(A,B,K,P,Y) PART 6∇;
COMMENT THIS PROCEDURE CAN THEN BE CALLED AS FOLLOWS;
INERPROD (A[I], B[I], ←5, I, INPRD);
```

COMMENT THIS IS EXACTLY EQUIVALENT TO THE CALL;
PART 6 $\{A=A[I]: B=B[I]; K\leftarrow 5; P=I; Y=\text{INPRD}\}$;
COMMENT TO USE THIS CODEPIECE AS A FUNCTION, WE CAN MODIFY STEP 6.4 AND DELETE STEP 6.5 BY TYPING:
6.4 RETURN (S);
DELETE STEP 6.5;
COMMENT THIS MODIFIED PART MAY NOW BE CALLED AS AN OPERAND;
$G\leftarrow H+$PART 6 $\{A=A[I]; B=B[I]; K\leftarrow 5; P=I\}-T$;

A conversation in LC[2] is initially in the hands of the user, but during the conversation either he or the system may be in control at various times. If the user gives control to his program, e.g., by a PART call, the program is responsible for the conversation. However, the user may intervene by interrupting the execution, in which case he becomes the initiator and may in fact start a new execution while holding the previous one in abeyance; thus conversations can be nested. The program may also initiate a conversation by executing a PAUSE statement or by executing a "?", a query mark which causes the system to request a value (or an expression to compute a value). For example, if the statement

1.1 $X\leftarrow ?Y+Z$;

were executed, the system would ask the user for a value for the variable Y. The query enabled by a "?" does not relinquish control of the conversation to the user as does a PAUSE; rather it demands a very specific reply from him—not simply any statement which he might wish to enter. In fact, if the user when asked for a value wants to gain control of the conversation, he must interrupt the system, temporarily suspending the program's query.

Another useful statement is the fill statement which occurs in two versions:

1.2 @ arbitrary comment string
  Halts execution, prints the arbitrary comment string, until the step 1.2 is entered from the console, after which it is executed. However, the next execution of 1.2 will require a repeat of the filling process.

1.2 @@ arbitrary comment string is the same as the above but the text entered is permanent and replaces the current statement 1.2.

It should be clear from the above that the man/system conversation is a recursive interaction, and a method of denesting a nested conversation must be available. This is done in two ways: if the program is in control, it "pops" the conversation back one level by completing its execution; the programmer can do the same by issuing a RESUME statement which gives conversational control back to the execution extant when he gained control. Since not all conversations are fruitful, a means of killing part or all of a nest of conversations (executions) is provided.

This is the EXIT statement whose function is to "close" parts already opened by previous calls. For example, if PART 1 had called PART 7 which in turn called PART 5, and PART 5 was executing when the user interrupted his program, he could decide to kill the execution back to PART 1 by typing:

EXIT TO PART 1;

This would cause parts 5 and 7 to be closed and would then return control to the user. The user is considered as a special PART called USER, and

EXIT TO PART USER;

will "pop" the conversation back to the last point at which the user had control. In this way, as many nested conversations as necessary may be discontinued along with their associated executions.

It is apparent that dialogue control is an important part of conversational systems. This dialogue control extends to the "apparent" control of LC^2 itself. Thus the user may define a PART 0 which intervenes between the user and LC^2.

Every input statement is passed to PART 0 whose execution is by LC^2. Naturally PART 0 can also control the execution of a user program. Thus without change of input syntax or with change of input syntax differential response of LC^2 to programmers is possible. To facilitate such action the user can access all control tables constructed by LC^2 during the execution of a user's program.

V. DATA STRUCTURES

All computer code sequences operate on data. The data is located through name and structure conventions which relate the data organization to its processing. Three tasks face the system designer:

The organization of the rules for data structuring.

The organization of the operations for data accessing, given a structuring.

The establishment of a useful syntax for the expression of data structures and the control of operations on them.

Two tasks face the programmer:

Mapping his own data into given structures.

Organizing the sequence of data accesses and alterations within the structures chosen.

It is in the nature of algorithms that their data are unbounded, though finite, and that detailed identification of unbounded data must be "indirect," i.e., not through programmer-chosen and predetermined names, which are inevitably bounded, but through the use of variables.

The processing of this data, as a fundamental preoccupation of the programmer, is controlled by the relative position of data within a structure or some other of its attributes. In the second case we say the data is unsorted. Unhappily, in general data cannot be simultaneously sorted on several attributes without making copies. If the data is "bulky," a copy may be made by merely pointing to the bulk, i.e., by using the name to stand for the object.

While many fundamental data organizations are possible, we limit our discussion to a few common and simple structures, each of which is but slightly different from the other:

1. The vector. The two basic properties of vectors are:
   (i) homogeneity: each component is of the same type or structure.
   (ii) index preservation: each component is attached to an unchanging index. Vectors may, however, have an increasing or decreasing number of components.
2. The string:
   (i) homogeneity
   (ii) nonindex preservation. The index attached to a component will not generally be fixed, i.e., it is not normal to manipulate strings by expecting to find a particular datum at the kth position. Inserts and deletions change the index of every element, e.g., following the position of change.
3. The list:
   (i) nonhomogeneity
   (ii) nonindex preservation

Since lists are nonhomogeneous, predicates for establishing the type of a datum are required. In LISP, e.g., the predicate *atom* serves this purpose. Lists may have more than one sequencing mode as a result of their nonhomogeneity. Thus in LISP, S expressions are "dotted pairs" $(X \cdot Y)$ and, for X and Y, lists, atoms, or the special terminal *nil* may be substituted. These lists are reasonable representations for trees and two-direction operators *car* (left) and *cdr* (right) are provided. Integer indexing is not in general a useful way to access data in a list.

List structures are possible, of course, with a variable number of components, none of which need be identical as to type.

4. The table:
   (i) index preservation
   (ii) nonhomogeneity

This is a weak definition since many tables are nonindex preserving, but consider it here as completing the classification.

A term often used in data processing is that of "file." A file can be any of the above, its chief characteristic being bulk and a data sequencing regime which places high penalties on nonmonotone processing, i.e., a file is processed in a monotone order based on some index. Note that the classification does not place any special emphasis on bounded data, this being regarded as somewhat irrelevant in a basic classification scheme.

It has been mentioned that copies may be indicated by pointers. This introduces a third classifier called self-referencing.

Any of the data structures may contain an additional datum called a pointer or reference which is a surrogate for the data pointed to. Thus, lists can be used to represent graphs, and strings to represent lists. Structures which can be self-referrent require an additional predicate to uncover pointers or references.

Every programming language has as basic types of data certain elementary or nonstructured "atoms," e.g., the character or the integer or the floating-point number, and one or more complex structures of the type mentioned. The complexes are chosen to best represent the processing rules programmers are likely to use when employing the language. In examining the widely used language systems, one sees that they are data dominated: the permitted data structures and their mode of use determine the syntax, the control structures, and even the supporting systems. ALGOL, while permitting recursion as a mode of control, is an iteration-dominated language. Its complex array is adequately controlled by iterations. Only when representing other structures in terms of arrays does recursive control prove of much value.

SNOBOL uses the string as its basic complex, the pattern as its basic predicate, and iteration as its basic control. While every substring of a string is a string, a single specific scan direction causes iteration to be the common control.

LISP, whose data structures are trees, uses recursion as the basic control and naturally so. The natural way to process a tree is by processing its subtrees, particularly since extensive substitution makes it tedious to control processing by counts on either depth or width. We have seen no language whose syntax so sublimely exposes the natural ways to process its data structure as that of LISP. However, we have yet to develop a language whose syntax is graceful in all of the common data structures.

Some languages and systems are being developed which permit the definition of data complexes by the programmer and the addition of operators in a new syntax to expedite the statement of tasks. These "extensible" languages will undoubtedly be heard from in the future.

It should be clear from the foregoing that there are major threads running through programming languages:

The binding of symbols
The organization of control regimes
The organization of data structures

The establishment of a "natural" syntax

Generally, languages overaccent some of these features to the detriment of others. Certainly, most programs seem to suit such an accenting. Nevertheless, a programming language system should provide all these features in some general fashion.

REFERENCES

Naur, P. (ed.), 1963. "Revised Report on the Algorithmic Language ALGOL 60," Comm. ACM 6, pp. 1–17.

Shaw, J. C. 1964. "JOSS: A Designer's View of An Experimental On-Line Computing System," *AFIPS PROCEEDINGS* **26**: 455–464.

Note: Section III, A is largely adapted from a preliminary report on LC² by J. G. Mitchell, A. J. Perlis, and H. VanZoeren which is to appear in the Proceedings of the ACM Symposium on Interactive Systems for Experimental Applied Mathematics, Academic Press, New York.